ENVIRONMENTAL TAXATION
AND THE DOUBLE DIVIDEND

CONTRIBUTIONS
TO
ECONOMIC ANALYSIS

246

Honorary Editor:
J. TINBERGEN†

Editors:
R. BLUNDELL
D. W. JORGENSON
J. -J. LAFFONT
T. PERSSON

ELSEVIER
Amsterdam – Lausanne – New York – Oxford – Shannon – Singapore – Tokyo

ENVIRONMENTAL TAXATION AND THE DOUBLE DIVIDEND

Ruud A. DE MOOIJ
Research Centre for Economic Policy (OCFEB),
Erasmus University Rotterdam, The Netherlands

CPB Netherlands Bureau for Economic Policy Analysis,
The Hague, The Netherlands

2000

ELSEVIER
Amsterdam – Lausanne – New York – Oxford – Shannon – Singapore – Tokyo

ELSEVIER SCIENCE B.V.
Sara Burgerhartstraat 25
P.O. Box 211, 1000 AE Amsterdam, The Netherlands

First edition 2000

Library of Congress Cataloging in Publication Data
A catalog record from the Library of Congress has been applied for.

ISBN: 0 444 50491 5

⊚ The paper used in this publication meets the requirements of ANSI/NISO Z39.48-1992 (Permanence of Paper).

Printed in The Netherlands.

INTRODUCTION TO THE SERIES

This series consists of a number of hitherto unpublished studies, which are introduced by the editors in the belief that they represent fresh contributions to economic science.

The term 'economic analysis' as used in the title of the series has been adopted because it covers both the activities of the theoretical economist and the research worker.

Although the analytical methods used by the various contributors are not the same, they are nevertheless conditioned by the common origin of their studies, namely theoretical problems encountered in practical research. Since for this reason, business cycle research and national accounting, research work on behalf of economic policy, and problems of planning are the main sources of the subjects dealt with, they necessarily determine the manner of approach adopted by the authors. Their methods tend to be 'practical' in the sense of not being too far remote from application to actual economic conditions. In addition they are quantitative.

It is the hope of the editors that the publication of these studies will help to stimulate the exchange of scientific information and to reinforce international cooperation in the field of economics.

The Editors

Preface

This book originates in the Dutch policy discussion in the beginning of the 1990s on the employment effects of environmental taxes. It was argued at that time that environmental tax reform, which substitutes pollution taxes for taxes on labor income, could serve two objectives, namely first a cleaner environment and second less labor-market distortions due to high taxes on labor income. In the Dutch policy discussion, environmental tax reform was generally viewed as a so-called double-edged sword. In the international literature, the idea that an environmental tax reform could kill two birds with one stone has become known under a different name, namely the double-dividend hypothesis. The research contained in this book throws cold water on the idea that environmental taxes are a cure for unemployment. By using simple analytical models and drawing on optimal tax theory, it points out that it is not evident at all that an environmental tax reform will raise employment by reducing tax distortions. The world is thus more complicated than the initial intuition about shifting the tax burden from labor to pollution would suggest.

Most of this research was carried out at the Research Centre for Economic Policy (OCFEB) at the Erasmus University Rotterdam. OCFEB aims to bridge the gap between academic research and the concerns of policymakers by pursuing policy-relevant research. The insights from this study have had a large impact on the discussion on environmental tax reform in the Netherlands. To illustrate, whereas at the beginning of the 1990s, the common wisdom was that an environmental tax reform could be an important instrument for raising employment, the last commission on green taxation in the Netherlands stated that environmental tax reform should be targeted primarily at reaping the first dividend, i.e. a cleaner environment; the second dividend does not appear to exist.

At the same time, the research contained in this book has made its mark also in the international academic literature. Earlier versions of some of the chapters in the book have been published in major journals such as the *American Economic Review*, the *European Journal of Political Economy*, the *Journal of Public Economics* and *International Tax and Public Finance*. This book reports on some of the most important and active research areas in environmental economics and public economics in the last decade. By studying environmental policy in the presence of non-environmental tax instruments, this research has revealed that environmental policy can interact importantly with these taxes. It has resulted in a careful rethinking of design of environmental policy in a world with pre-existing distortions. More generally, it has given a new impetus to the investigation of how economic policy interacts with pre-existing distortions in the economy. Thus, this book is an excellent example of what OCFEB stands for: the pursuit of high quality research that is relevant for actual policymaking.

After surveying the literature, the book provides a rigorous analysis of a benchmark model. This model shows how environmental levies may exacerbate labor-market distortions due to a distortionary tax on labor income. The basic framework of the benchmark model is subsequently enriched to deal with various extensions: the capital market and international capital mobility, international trade, distributional considerations, imperfect labor markets and involuntary unemployment, feedbacks of environmental quality on the economy, and endogenous growth. By taking the same benchmark model as a starting point, the separate effects of the various extensions become clear. The book illustrates the great force of small analytical models to uncover the main transmission channels of environmental taxation. All chapters (except chapter 9) also contain numerical simulations to explore the magnitude of the effects.

Many countries are currently considering and implementing environmental tax reforms. The failure of the double-dividend hypothesis as documented in this book does not imply that these countries are misguided. Whereas the second dividend may be doubted, the first dividend (i.e. a cleaner environment) remains a powerful reason for the introduction of pollution taxes. Indeed, the main lesson of this study is that the desirability of pollution taxes depends on the environmental benefits, i.e. the degree to which it serves the first dividend. It therefore points to the importance of documenting the environmental benefits. The double dividend is dead. Long live environmental taxation!

Lans Bovenberg

Contents

1   Introduction

Should governments impose higher taxes on "bads," such as polluting activities, and lower taxes on "goods," such as work, investment and consumption? Many people feel sympathetic to this idea. Indeed, environmental taxes seem attractive for a number of reasons. First of all, environmental taxes accord with the polluter-pays principle, which is subscribed to by all OECD countries. Second, it is widely believed that, by raising the price of pollution, environmental taxes put a brake on the activities that harm our natural environment. In this way, such taxes correct for market failures by internalizing the external effects of environmentally damaging activities. Third, compared to other environmental policy instruments, environmental taxes are attractive because they are often able to achieve a certain environmental objective in a cost-effective way (Baumol and Oates, 1988). Finally, environmental taxes are sometimes desirable on other grounds such as compliance, administration or political viability (Fullerton, 1995).

For these reasons, environmental taxes have received increasing attention during the last decade. One important example of their popularity is the international debate on carbon taxes to combat global warming. In particular, at the Kyoto conference in 1997, industrialized countries committed themselves to reduce CO_2 emissions. For instance, compared to the emission level in 1990, the European Union will have to reduce carbon emissions by 8% in 2010. A carbon tax is often considered a promising policy option to meet such a target (EC, 1997). Also the OECD has investigated the possibilities for a uniform carbon tax in all OECD countries as a policy of climate protection (OECD, 1997). Some Northern European countries have recently imposed such environmental taxes on a unilateral basis (Ekins and Speck, 1997).

Some economists have suggested that environmental taxes are attractive not only for the reasons mentioned above, but also because these taxes raise public revenues. Currently, public revenues in industrialized countries are raised primarily through taxes on income and consumption, while environmental taxes yield only small revenues. However, whereas income taxes *increase* distortions, e.g. by discouraging labor supply and investment, environmental taxes *alleviate* distortions by discouraging pollution. It has therefore been suggested to shift the tax burden away from distortionary taxes towards environmental taxes, thereby reducing the overall efficiency cost of the tax system. Indeed, Pearce (1991) and Repetto et al. (1992) have presented estimates for the deadweight loss of current tax systems to illustrate the magnitude of the potential efficiency gains from such an environmental tax reform. It is thus argued that governments may reap the twin objectives of improving environmental quality and reducing the excess burden from taxation. This is the so-called 'double-dividend hypothesis.'

In recent years, policy makers and environmental interest groups have shown great interest in the double dividend. For instance, environmentalists usually had to justify tough environmental policies on the basis of environmental benefits, which are often surrounded by considerable uncertainty.[1] The double-dividend claim suggests that environmental improvements come free. Accordingly, environmentalists would no longer have to defend increases in environmental taxes on the basis of their uncertain environmental benefits, but could rely instead on the non-environmental benefits, such as increases in income or employment. Moreover, other policy makers believed that the double dividend would make life easier. For instance, European policy makers sometimes perceived the double dividend as a way to tackle the high structural unemployment levels, without imposing painful reforms of the welfare states. Indeed, discussions on environmental taxes in a number of EU countries in the early 1990s focused primarily on the magnitude of the second, non-environmental dividend, rather than on the expected environmental benefits.

---

[1] See, for instance, the on-going debate among scientists about the consequences for the global climate of higher concentrations of greenhouse gases in the atmosphere.

Since the early 1990s, several economists have investigated the validity of the double-dividend hypothesis. In particular, they have been exploring whether an environmental tax reform is able to yield both an environmental dividend (i.e. environmental benefits due to less pollution) and a second, non-environmental dividend.[2] In various studies, economists have used several definitions and interpretations of the non-environmental dividend. In this book, I distinguish three forms:[3]

1. The *weak double dividend* states that, by using the revenues from environmental taxes to cut pre-existing distortionary taxes, one can achieve cost savings compared to the case where the revenues of the environmental taxes are recycled in a lump-sum fashion. By cost savings, I mean reductions in the excess burden from taxation viewed from a non-environmental perspective (i.e. not considering the environmental benefits).

2. The *strong double dividend* holds if, compared to the initial situation, an environmental tax reform raises welfare through both environmental benefits and improvements in the efficiency of the tax system viewed from a non-environmental perspective.[4]

3. The *employment double dividend* is obtained if, compared to the initial situation, an environmental tax reform both improves the quality of the environment and boosts employment.

---

[2] A separation into two dividends is possible only if there are no feedbacks of environmental quality on the economy. If improvements in environmental quality exert feedback effects, e.g. by affecting productivity, then the two dividends partly coincide (see chapter 8).

[3] The first two forms are due to Goulder (1995a), while the third is used by Carraro et al. (1996).

[4] In this book, the second dividend is sometimes referred to as the efficiency (cost) of the tax system. This narrow definition of efficiency (cost) does not include environmental benefits, which are captured by the first dividend.

Whereas the weak double dividend compares the effects of an environmental tax reform with the hypothetical case of lump-sum revenue recycling, the two other forms of the double dividend compare the situation before and after the reform. According to Goulder (1995a), the weak double dividend is relatively uncontroversial. Most theoretical contributions to the literature therefore deal with the strong double dividend. In Europe, a number of empirical studies focus on the employment double dividend.[5]

This book fits into the literature on the double dividend. Moreover, it is related to the literature on optimal pollution taxation. The latter studies generally explore how the second-best optimal pollution tax deviates from its first-best optimum. In particular, in a first-best world without other distortions, it is well-known that governments find it optimal to fully internalize environmental externalities. This first-best optimum can be achieved by setting environmental taxes equal to the marginal environmental damage from polluting activities. At this so-called Pigovian tax, a marginal change in the pollution tax does not affect welfare through the environmental distortion since the marginal social costs and benefits of pollution coincide. In a second-best world, however, environmental taxes may impact welfare through a second, non-environmental distortion. Accordingly, the optimal second-best pollution tax typically differs from its first-best rate.

Chapter 3 of this book investigates the welfare effects of environmental taxes in a simple second-best framework with environmental externalities and prior labor-tax distortions. Within this framework, I derive also the optimal second-best pollution tax. Subsequent chapters extend this so-called *benchmark model* in several directions. In this way, the book contributes to the literature on the double dividend and optimal second-best pollution taxes in three important

---

[5] The environmental taxes discussed in this book are general in nature and can be interpreted as taxes on any type of polluting activity, including resource waste, pollution, and congestion. In recent policy debates, taxes on energy or carbon have received the most attention. This example might serve best to interpret the results. In the simulations, I adopt energy data to numerically explore the effects of environmental taxes.

ways. First, it provides a comprehensive overview of the theoretical literature in a formalized framework. Most survey papers on the double dividend give an intuitive explanation of the main findings, but not a formalization.[6] This is useful since economic intuition alone is not always a reliable guide if one investigates problems in a second-best world. A second contribution of this book is that, by analyzing different extensions of the benchmark model, it explores whether the main findings from the simple theoretical model carry over to more complex frameworks that are closer descriptions of the real world. In particular, the extensions indicate that policy makers face several trade-offs when designing environmental tax reforms. Finally, by performing simulations with the different theoretical models, the book tries to bridge the gap between the intuition from theoretical studies and the results from empirical models on the employment double dividend. On the one hand, the stylized theoretical models often ignore several relevant interactions that actually take place in the economy. On the other hand, the applied models are sometimes quite complicated -- so that the main driving forces behind the results are unclear. By combining advanced theoretical models with empirical information on the key parameters, this book tries to discover the transmission channels that drive the results in empirical models. In this way, the analysis helps to shed light on the simulations from applied research.

Chapter 2 of this book provides a survey of the double-dividend literature. It starts with a brief overview of some early theoretical studies on second-best pollution taxes and elaborates on the origin of the double-dividend literature in the early 1990s. It then discusses the intuition behind the main findings from theoretical and empirical studies on the double dividend and optimal second-best pollution taxes. Chapter 2 discusses also some political-economy aspects related to the double-dividend.

Chapter 3 develops a small general equilibrium model -- the so-called benchmark model -- to analyze the welfare effects of an environmental tax

---

[6] An exception is Bovenberg (1997).

reform in a second-best framework. The model describes a small open economy that faces exogenous world-market prices for commodities. Labor is immobile internationally so that the wage rate is determined endogenously. Firms produce output by combining labor inputs and other inputs that pollute the environment. Households receive only labor income -- which is spent on two consumption goods, one of which adversely affects the environment as it is consumed. Taxes are assessed on labor, the polluting input and the polluting consumption good.

The simple model shows that environmental taxes exacerbate, rather than alleviate, pre-existing tax distortions, even if the revenues are used to cut taxes on labor. Accordingly, the optimal second-best pollution tax is smaller than the Pigovian rate. Furthermore, a revenue-neutral reform from labor towards either polluting inputs or polluting consumption does not yield a strong double dividend. The intuition behind these results is that the incidence of pollution taxes ultimately falls on labor. Hence, an environmental tax reform boils down to a replacement of an explicit labor tax for an implicit labor tax. The implicit labor tax is typically less efficient as an instrument to raise public revenues than the explicit labor tax because it features a narrower tax base.[7] Indeed, the narrow-based tax induces economic agents to avoid polluting behavior so that environmental quality improves. At the same time, however, this behavioral response erodes the tax base, thereby raising the efficiency cost of taxation from a non-environmental point of view. Intuitively, the environmental tax raises the supply of the public good of the environment. The abatement costs necessary to improve the quality of the environment are borne by households in the form of lower private incomes. I call these private costs associated with a cleaner environment the *tax burden effect*.

Chapter 4 extends the analysis of chapter 3 by incorporating capital in the production function. This allows for an inefficient distribution of the tax burden over two non-polluting production factors (i.e. labor and capital). Within such a setting, an environmental tax reform induces not only a tax burden effect, but

---

[7] This result depends on the assumptions of homothetic utility and leisure being weakly separable from private consumption goods.

also a so-called *tax shifting effect*. In particular, the reform may shift the tax burden from the factor that is relatively overtaxed from a non-environmental point of view towards the factor that is relatively undertaxed. In that case, the tax shifting effect alleviates initial inefficiencies in the tax system so that a double dividend may become feasible. Whether the tax shifting effect indeed works in the right direction depends importantly on the degree of capital mobility. In particular, with rather immobile capital, a double dividend is feasible if an environmental tax reform shifts the tax burden from labor towards capital. However, in the long run capital is rather mobile. In that case, tax shifting towards capital may exacerbate initial inefficiencies associated with capital taxation.

In chapter 5, the home country may exert monopsony power on the international market for polluting inputs. In that case, pollution taxes operate as implicit tariffs. Indeed, environmental levies can change the terms of trade in favor of the home country. Accordingly, the home country shifts the tax burden towards foreign suppliers of polluting inputs, thereby raising domestic non-environmental welfare. If international agreements preclude explicit tariffs to be adopted for trade policy, environmental taxes may thus be employed as implicit trade-policy instruments aimed at improving the terms of trade.

Simulations with the model suggest that terms-of-trade effects are potentially important mechanisms in case of coordinated CO_2 policies. Indeed, for plausible parameter values, chapter 5 finds that an OECD-wide energy tax reform may yield a double dividend due to favorable terms-of-trade effects. However, if the reform is implemented on a smaller scale, such as the EU, the potential for a double dividend declines.

Chapter 6 modifies the household model from chapter 3 by introducing heterogenous agents. In particular, households differ with respect to their skill level -- which causes also differences in the labor income they receive. Society may feature preferences for an equitable distribution of welfare among the different households. This can be achieved through the provision of uniform lump-sum transfers or an allowance for tax payments on polluting commodities.

Chapter 6 finds that an environmental tax reform may produce a double dividend by improving the quality of the environment and reducing the efficiency cost of taxation (where efficiency does not include the social value attached to distributional concerns). However, such a double dividend is feasible only if the tax burden is shifted away from households that receive relatively high labor incomes towards households that rely more heavily on transfer incomes. Accordingly, despite the presence of a double dividend, social welfare may decline if society features strong preferences for equity. Alternative options for recycling the revenues from the environmental tax may yield more favorable outcomes from an equity point of view, but these reforms fail to reduce the efficiency cost of taxation.

Chapter 7 develops a trade-union model to describe the process of wage formation in an imperfect labor market. In this framework, an environmental tax reform may raise employment if it hurts the fall-back position of the union, relative to the position of workers. Intuitively, a less attractive fall-back position reduces the bargaining power of the union in wage negotiations. This moderates wages and stimulates employment. Whether an environmental tax reform succeeds in reducing the fall-back position of the union depends on how income during unemployment is indexed to wages. In particular, if the unemployed collect incomes that are not subject to labor taxes, but rather are taxed by the environmental charges (such as incomes from informal activities), an environmental tax reform reduces the bargaining position of the union. However, if income during unemployment is indexed to real after-tax wages, an environmental tax reform has no implications for their bargaining position. In that case, such a reform may nevertheless raise employment if pollution taxes bear more heavily on profits than they do on labor. Empirical evidence on wage formation suggests that lower labor taxes indeed moderate wage costs in a number of EU countries. This explains the favorable outcome for the employment double dividend that is sometimes found by empirical models.

Chapter 8 addresses the issue of feedback effects from environmental quality on the economy. First, it introduces environmental externalities in production.

Second, it allows for feedback effects of environmental quality on private behavior. In the presence of feedback effects from environmental quality on the economy, the environmental dividend and the non-environmental dividend partly coincide. Accordingly, the two dividends can no longer be separated in a straightforward manner. In chapter 8, I attribute the feedbacks of environmental quality on the economy to the second, non-environmental dividend. In that case, the analysis reveals that non-environmental welfare may rise due to an environmental tax reform in two cases: first, if production externalities are sufficiently large and, second, if environmental quality is a sufficiently strong substitute for leisure in the utility function.

Chapter 9 extends the double-dividend analysis by employing a dynamic model of endogenous growth. Whether an environmental tax reform produces a non-environmental dividend depends on how such a reform impacts the growth rate. In particular, by driving a wedge between the marginal social costs and benefits of capital accumulation, distortionary income taxes reduce economic growth below its first-best level. Accordingly, a reform that boosts growth alleviates the deadweight loss from the distortionary income tax.

Chapter 9 uncovers two channels through which an environmental tax reform may yield a double dividend. The first channel is -- similar to chapter 8 -- a positive environmental externality in production associated with the role of the environment as a public production factor. The second channel is a shift in the tax burden away from the return on capital accumulation towards profits. In particular, a pollution tax may improve the efficiency of the tax system as a revenue-raising device by taxing away the quasi-rents from pollution. This second channel operates only if substitution between pollution and other inputs is difficult, so that the base of the pollution tax is rather inelastic.

2   A survey of the double-dividend literature

The literature on the double dividend deals with the welfare effects of environmental taxes in a world with pre-existing tax distortions. This chapter provides a survey of this literature. In particular, it discusses the literature on both environmental tax reform and optimal second-best pollution taxes. The introductions in chapters 4 to 9 provide more detailed reviews of the literature on specific aspects of the double dividend and optimal second-best pollution taxes such as the relationship with capital taxation, terms-of-trade effects, distributional concerns, imperfect labor markets, environment-economy interactions and economic growth.

Several other survey papers on the double dividend can be found in the literature. Goulder (1995a) clarified the debate by distinguishing between different forms of the double dividend. In particular, he explores whether these different forms are confirmed by theoretical and empirical studies. Oates (1995) provides the intuition behind the main findings in the double-dividend literature and briefly discusses some issues related to the actual design and implementation of pollution taxes. Bovenberg (1995) presents a general assessment of the double dividend in a paper with a strong policy focus. In particular, he discusses several real-world phenomena that may influence the main conclusions. De Mooij and Vollebergh (1995) start from fundamental tax principles to put the double dividend in a broader perspective of tax design. Fullerton and Metcalf (1997) provide a detailed historical review of the literature on second-best pollution taxes. They raise the question whether the emphasis of the double-dividend literature on the revenues from environmental taxes is justified. Bovenberg (1997) surveys the literature on the basis of an analytical general equilibrium model, thereby elaborating on the key parameters and economic mechanisms involved. In Bovenberg (1998), he reflects on the political economy of environmental tax reforms. Pezzey and Park (1998) put the double dividend in a broader context of instrument choice.

    This chapter is organized as follows. Section 2.1 discusses the history of the double dividend. Section 2.2 elaborates on the main findings from recent theoretical studies that suggest that the double dividend typically fails. Section 2.3 explores some channels through which a double dividend may become feasible. In section 2.4, some political-economy issues are discussed, while section 2.5 elaborates on the relevance of environment-economy interactions. Finally, section 2.6 concludes.

2.1   Early literature

In his seminal book, Pigou (1932) showed that pollution taxes are able to internalize the adverse external effects associated with polluting activities. Indeed, in a first-best world without other distortions, it is optimal to set the pollution tax equal to the marginal environmental damage from pollution. Most of the literature after Pigou has focused on this corrective task of environmental taxes, thereby largely ignoring the revenues raised by these taxes. Indeed, most studies have assumed that the revenues raised by the so-called Pigovian taxes are returned to economic agents in a lump-sum fashion. Furthermore, most studies have assumed that the environmental distortion is the only market failure, thereby ignoring the presence of other tax or non-tax distortions in the economy.

2.1.1   Environmental tax reform

Tullock (1967) was the first to recognize the significance of the revenues from environmental taxes. In particular, he argues that government revenue might be 'free money' if raised from charges on an environmental externality. Tullock thus recognized that public revenues are scarce if we live in a world where lump-sum taxes are not available. Terkla (1984) more explicitly investigated the importance of the revenue-raising ability of environmental taxes. Using a numerical partial equilibrium model, he finds that raising pollution taxes and using the revenues to cut other distortionary taxes -- rather than recycling them in a lump-sum fashion as in the world of Pigou -- reduces the excess burden from taxation. In fact, Terkla thus supports the weak double-dividend claim because he compares a reduction in distortionary taxes with the case of lump-sum revenue recycling.

The significance of the revenues from environmental taxes has been the starting point for the double-dividend literature. In particular, Pearce (1991) introduced the term "double dividend" in his paper on carbon taxes as an instrument to combat global warming. Pearce states: "Governments may then adopt a fiscally neutral stance on the carbon tax, using revenues to finance reductions in incentive-distorting taxes such as income tax, or corporation tax. This 'double-dividend' feature of a pollution tax is of critical importance in the political debate about the means of securing a carbon convention." (p. 940). In contrast to Terkla, Pearce seems to have the strong double dividend in mind. Repetto et al. (1992) more explicitly focus on the strong double dividend by arguing that, whereas taxes on income and profits distort economic decisions, environmental taxes correct existing distortions related to damaging activities. These authors thus argue that shifting away from ordinary taxes towards environmental charges could yield substantial efficiency gains without loss of public revenues.

Other studies that emphasize the revenue-raising capacity of environmental taxes have confirmed the weak double dividend. To illustrate, Ballard and Medema (1993) adopt a numerical simulation model to analyze the differences in welfare effects between a tax on pollution and a subsidy on pollution abatement. They find that pollution taxes are more efficient than subsidies because the revenues of pollution taxes can be used to reduce other taxes that impose large excess burdens on the private sector. Also Repetto and Austin (1997) confirm the weak double dividend by investigating simulations from 16 large-scale economic models. They find that, compared to lump-sum revenue recycling, more efficient ways of revenue recycling significantly reduce the predicted cost of a carbon tax.

2.1.2   Tax reform versus optimal tax

Optimal tax literature in the presence of externalities generally focuses on how the optimal second-best pollution tax differs from the Pigovian rate. Although the double dividend originates in the literature on environmental tax reform, it is closely related to this literature on optimal second-best pollution taxes.[8] In particular, if a (marginal) environmental tax reform raises welfare on account of both an environmental dividend and a non-environmental dividend, it moves the tax system closer to the optimum. Hence, if the strong double dividend holds true when we start from the Pigovian rate, this indicates that the optimal second-best pollution tax exceeds this first-best Pigovian level. However, the strong double dividend imposes a sufficient, but not a necessary condition. Indeed, even if the strong double dividend fails at the Pigovian rate, the second-best optimal tax may still exceed the first-best level.

This is illustrated by following an alternative approach to tax-reform analysis. In particular, the welfare effects of an environmental tax reform can be divided into the marginal effects on two distortions -- instead of two dividends -- namely the environmental and the non-environmental distortion. The environmental distortion is measured by the difference between the marginal social costs and benefits from additional pollution. This contrasts with the environmental dividend discussed above, which is measured by the marginal social costs of additional pollution alone. If the government sets the pollution tax equal to the first-best Pigovian rate, the marginal social costs and benefits of pollution coincide so that the environmental distortion is zero. At the Pigovian rate, a marginal increase in the pollution tax thus does not affect welfare through the channel of the environmental distortion. However, if we are in a second-best framework, such a marginal tax reform may affect welfare through the channel of a non-environmental distortion. Therefore, the optimal second-best pollution tax typically deviates from its Pigovian level. In particular, if a marginal increase

---

[8] Part of the literature on optimal second-best pollution taxes considers monopoly power as the non-environmental distortion (see e.g. Buchanan, 1969; Barnett, 1980). This book, however, concentrates on pre-existing tax distortions.

in the environmental tax above the Pigovian level improves overall welfare by alleviating the non-environmental distortion, the second-best optimal pollution tax exceeds the Pigovian rate. In contrast, if the environmental tax reform exacerbates the initial non-environmental tax distortion, the second-best optimal tax is smaller than the Pigovian rate. Accordingly, this imposes a necessary condition for the second-best optimal pollution tax to exceed the first-best level: a marginal environmental tax reform away from the Pigovian rate should alleviate the non-environmental distortion in the economy. This condition typically differs from the one for a strong double dividend. Indeed, Bovenberg and de Mooij (1994b) show that the conditions for a strong double dividend are more stringent than the conditions for the alleviation of the non-environmental tax distortion (see also chapter 3 of this book).

Vice versa, optimal-tax analysis may also yield information about the welfare effects of an environmental tax reform. In particular, if the optimal second-best pollution tax exceeds the Pigovian level, a marginal increase in the pollution tax above the Pigovian tax alleviates the pre-existing non-environmental distortion in the economy. However, it does not mean that a marginal increase in the pollution tax above the Pigovian level yields a strong double dividend.

2.1.3   Optimal second-best pollution taxes

Sandmo (1975) was the first to derive the optimal tax rate on an externality-generating good in a second-best framework with distortionary taxes. Using a simple general equilibrium framework, Sandmo derives the following optimal tax rate on an externality-generating good as part of an optimal commodity tax system:[9]

---

[9] Ng (1980) performed a similar exercise, but explored (the sign of) the optimal tax on externality-generating goods in the presence of distortionary income taxes, rather than in the presence of commodity taxes.

$$T^* = (1 - \frac{1}{\eta})T^R + \frac{1}{\eta}T^P \tag{2.1}$$

Expression (2.1) reveals that the optimal environmental tax, T^*, is determined by two terms, reflecting the two goals that the government wants to accomplish with the environmental tax. The first term on the right-hand side (RHS) of (2.1) is the revenue-raising component, T^R, which denotes the revenue-raising task of environmental taxes. In a partial equilibrium framework, this term would involve the inverse-elasticity formula. More generally, it reflects the well-known Ramsey tax rule. The second term on the RHS of (2.1) represents the Pigovian component, T^P, which is related to the externality-correcting task of environmental taxes. The term η stands for the so-called marginal cost of public funds (MCPF) and measures how scarce public funds are relative to private incomes. In particular, if the government requires distortionary taxes to raise revenues, public funds are scarcer than private funds -- so that the MCPF lies above unity. The result from Sandmo in (2.1) reveals that the optimal second-best pollution tax generally deviates from its first-best level (equal to T^P). However, it does not give us an unequivocal answer to the question whether the optimal pollution tax is below or above the Pigovian tax.

In contrast to this, Lee and Misiolek (1986) focus explicitly on the magnitude of the optimal second-best pollution tax. Using a partial equilibrium model, they suggest that, in the presence of other distortionary taxes, the optimal pollution tax exceeds the Pigovian rate as long as the pollution tax raises a positive amount of public revenue. This claim suggests that, starting from the Pigovian rate, the government alleviates pre-existing tax distortions by marginally increasing environmental taxes and reducing other distortionary taxes. Also Oates (1993) stresses that the revenue effect of pollution taxes will have implications for the optimal tax on pollution. However, he maintains that this should be investigated in a general equilibrium framework. Nordhaus (1993) illustrates the findings from Lee and Misiolek numerically. On the basis of calculations with the DICE model, he concludes that the optimal carbon tax in the US rises from \$5 per ton in a first-best setting to \$59 per ton in a second-best world with pre-existing tax distortions.

2.2   Recent literature

The studies by Pearce (1991) and Repetto et al. (1992) suggest that substituting pollution taxes for ordinary taxes on income may both improve environmental quality and, at the same time, raise welfare from a non-environmental point of view (i.e. it yields a strong double dividend). Lee and Misiolek (1986) and Nordhaus (1993) argue that the optimal second-best pollution tax is above the Pigovian level. Hence, a marginal increase away from the Pigovian rate will alleviate pre-existing non-environmental tax distortions. This section discusses the recent literature, which reveals how the intuition behind these results might go wrong.

2.2.1   Environmental tax reform

Using an analytical general equilibrium model, Bovenberg and de Mooij (1994a) have analyzed the welfare effects of a reform from distortionary labor taxes towards environmental taxes. They find that environmental taxes typically reduce employment and thus exacerbate pre-existing tax distortions on the labor market, even if the revenues from the environmental taxes are used to reduce the labor tax. Bovenberg and de Mooij thus conclude that the optimal second-best pollution tax is typically smaller than its first-best level. According to Bovenberg and de Mooij (1994b), this implies also that the strong double dividend fails (see also chapter 3). Hence, there exists a trade-off between improvements in environmental quality and reductions in the excess burden from taxation.

The intuition behind these results is that an environmental tax reform involves a shift from a broad-based tax on income towards a narrow-based tax on pollution. These narrow-based taxes are typically less efficient than broad-based taxes, since they induce more behavioral responses. Indeed, narrow-based environmental taxes aim at encouraging substitution from polluting behavior towards activities that are less damaging for the environment. Although good for the environment, these behavioral responses reduce tax revenues. The government thus needs to raise other taxes to maintain the government budget -- which exacerbates pre-existing tax distortions on the labor market and reduces private incomes. Hence, the objective to encourage environmentally-friendly behavior (the environmental dividend) conflicts with the aim of the tax system to raise public revenues with the lowest costs to private incomes (the non-environmental dividend).[10]

An alternative interpretation of the failure of the strong double dividend is the following. By enhancing environmental quality, pollution taxes raise the supply of the public good of the environment. In this way, environmental taxes raise the overall tax burden associated with government intervention. The abatement costs necessary for the improvement in environmental quality are borne privately in the form of lower private incomes. This discourages labor supply, thereby exacerbating the pre-existing distortions on the labor market. Indeed, whereas the costs of pollution taxes are borne privately and reduce labor supply, the benefits of pollution taxes in the form of a cleaner environment are public and thus leave labor supply unchanged. The efficiency cost associated with a more distortionary tax system is called the *tax burden effect* of an environmental tax reform.[11]

In the US, empirical macroeconomic models have been employed to investigate whether an environmental tax reform may yield a strong double

---

[10] If polluting commodities would be a relative complement to leisure, these behavioral responses may raise efficiency, since pollution taxes discourage the consumption of leisure. Accordingly, it would be optimal for efficiency reasons to tax polluting commodities more heavily than other commodities (Corlett and Hague, 1953; Ng, 1980; Parry, 1995; Fuest and Huber, 1999). Bovenberg and de Mooij (1994ab) assume that clean and polluting commodities are equally good substitutes for leisure. In that case, uniform taxes on clean and polluting commodities are optimal from a non-environmental point of view. For taxes on polluting inputs, Diamond and Mirrlees (1971) have shown that aggregate production efficiency requires zero taxes on intermediate inputs.

[11] Bovenberg and van der Ploeg (1994ab) and van der Ploeg and Bovenberg (1994) explore also the optimal choice between different public priorities, including environmental quality and public consumption.

dividend. These studies have focused primarily on the welfare costs of carbon taxes, recycled through lower income taxes. Goulder (1995a) concludes, on the basis of a survey of 11 studies, that the strong double dividend typically fails in empirical simulation models for the US.

2.2.2   Optimal environmental tax

In Bovenberg and de Mooij (1994a), the non-environmental distortion originates in a positive initial labor tax, which drives a wedge between the marginal social costs and benefits of an additional unit of employment. Accordingly, a necessary condition for the optimal second-best pollution tax to exceed the Pigovian rate is that a marginal increase in the pollution tax above the Pigovian rate increases employment. As employment falls in the model of Bovenberg and de Mooij, the optimal second-best pollution tax is below the first-best level.[12]

Bovenberg and van der Ploeg (1994a) and Parry (1995) have used expression (2.1) to clarify discussions on the optimal second-best pollution tax. In particular, expression (2.1) reveals that the weights that the government assigns to its revenue-raising task and the environmental task depend on the MCPF: a more distortionary tax system -- as reflected by a higher MCPF -- increases the weight of the Ramsey component and reduces the weight of the Pigovian component. Hence, a more distorted tax system makes it less attractive for the government to use taxes aimed at correcting externalities and more urgent to use taxes that raise revenues with lowest cost to private incomes. Intuitively, pollution taxes that aim at reducing pollution are typically more distortionary instruments than other taxes in terms of revenue raising. A large MCPF magnifies these efficiency costs of pollution taxes.

---

[12] Håkonsen (1998) argues that this issue has been misinterpreted by Goulder (1995a). In particular, Goulder suggests that a sufficient condition for a strong double dividend in Bovenberg and de Mooij's model is a negative uncompensated wage elasticity of labor supply, which ensures that employment expands on account of an environmental tax reform (see Håkonsen, 1998, p. 53). As indicated above, this condition is necessary but not sufficient for a strong double dividend.

For a particular normalization of the tax system, Bovenberg and van der Ploeg (1994a) show that the Ramsey component in (2.1) is zero. Accordingly, the optimal pollution tax is below the Pigovian rate if the MCPF exceeds unity. As pointed out by Schöb (1997) and Fullerton (1997), the Ramsey component in (2.1) may be positive for alternative normalizations. In that case, the optimal pollution tax may exceed its Pigovian rate. It is therefore more general to state that the optimal tax difference between the polluting good and other goods is less than the Pigovian rule suggests. This result is related to the literature on the optimal provision of public goods. In particular, the modified Samuelson rule suggests that the marginal rate of substitution between public and private consumption should exceed the marginal rate of transformation if the MCPF exceeds unity (Atkinson and Stern, 1974). In the context of the supply of the public good of the environment, a high MCPF reduces the optimal tax difference between clean and polluting goods below the Pigovian rate.[13]

Parry (1995) provides an intuitive explanation for the optimal-tax result of Bovenberg and de Mooij and Bovenberg and van der Ploeg. He argues that, compared to the first-best world, the effects of an environmental tax reform in a second-best framework are modified in two directions. First, the revenues from environmental taxes are not necessarily recycled in a lump-sum fashion but can be used to cut pre-existing distortionary taxes. This might enhance the efficiency of the tax system as a revenue-raising instrument. This first modification was emphasized also by the proponents of the double dividend. Parry calls this the revenue effect.[14] A second modification is that the efficiency cost of pollution taxes are typically magnified through general equilibrium interactions. Indeed, environmental taxes affect not only the market for polluting commodities, but also other markets like the labor market. To illustrate, pollution taxes on households not only induce substitution from polluting consumption towards clean consumption, but also reduce the real wage. This discourages the

---

[13] This does not mean that environmental quality in a second-best world is worse than in a first-best setting (see Gaube, 1998). Indeed, pre-existing tax distortions may already internalize part of the environmental externality, e.g. by reducing economic activity.

[14] Goulder (1995a) refers to this as the revenue-recycling effect.

incentives to supply labor, which exacerbates pre-existing tax distortions on the labor market. This is called the tax-interdependency effect (or tax-interaction effect). As pointed out by Parry, the tax-interaction effect is typically larger in magnitude than the revenue-recycling effect under plausible assumptions. Accordingly, the optimal second-best pollution tax is smaller than the Pigovian tax. The partial equilibrium models employed in the early literature on second-best pollution taxes did not recognize the tax-interaction effect. This explains the discrepancy between the claims from earlier studies on the double dividend and the more recent results from general equilibrium models.

The optimal-tax result of Bovenberg and van der Ploeg is consistent with calculations of the optimal second-best pollution tax by Bovenberg and Goulder (1996). Employing a dynamic CGE model for the US, they find that the optimal carbon tax is approximately 68% of the Pigovian rate.[15] Using a somewhat simpler general equilibrium framework, Parry (1995) also finds that the optimal pollution tax in the US lies between 63% and 78% of the Pigovian rate. These optimal-tax results contrast with the findings from Lee and Misiolek (1986) and Nordhaus (1993), who suggest that the optimal pollution tax exceeds the Pigovian tax. The explanation for this discrepancy in results is that these latter authors have ignored general equilibrium interactions between environmental taxes and pre-existing tax distortions.

2.3   Tax-shifting

The analytical results from Bovenberg and de Mooij (1994ab), Bovenberg and van der Ploeg (1994a) and Parry (1995) may not be very general. Indeed, although their second-best frameworks are an important step closer to what we observe in the real world than Pigou's first-best setting, the results are derived from stylized analytical models. These contain a number of simplifying assumptions that are not consistent with reality. Therefore, economists have

---

[15] Bovenberg and Goulder find that the optimal carbon tax would be somewhat higher if the rest of the tax system would be closer to its non-environmental optimum. Hence, pre-existing inefficiencies in the US tax system further magnify the efficiency costs of carbon taxes.

explored second-best pollution taxes in more realistic settings by extending the stylized models in several directions. In general, these extensions reveal that environmental tax reforms may affect the efficiency costs of taxation through two channels. First, they produce a tax burden effect that is responsible for the failure of the strong double dividend in the stylized models discussed above. Second, if the initial tax structure is sub-optimal from a non-environmental point of view, pollution taxes may improve the overall efficiency of taxation by moving the tax system closer to its non-environmental optimum. This is called the *tax-shifting effect*. Below, we discuss three types of tax-shifting effects that have received attention in the literature.

2.3.1   Tax shifting between factors

The first extension of the stylized model that allows for a tax-shifting effect is the modeling of more inputs in production (see chapter 4). In that case, the initial tax structure over different inputs may be sub-optimal from a non-environmental point of view. To illustrate, if the marginal excess burden of labor taxes exceeds that of capital taxes, labor is overtaxed compared to capital in the initial equilibrium. Hence, a shift in the tax burden from labor towards capital would reduce the efficiency cost of the tax system. Political constraints may preclude such a direct shift in the tax system, however. In that case, a reform from labor taxes towards pollution taxes may be used as an indirect way to shift the tax burden across labor and capital. In particular, the burden of environmental taxes falls primarily on capital if capital and pollution are relative complements in production and polluting inputs are supplied elastically. In that case, an environmental tax reform can move the tax system closer to its non-environmental optimum so that the second dividend becomes feasible.

Whether tax shifting to capital is likely to occur depends -- apart from the degree of complementarity between capital and polluting inputs -- on the mobility of capital. In the short run, capital is rather immobile so that capital incomes may share in the burden of taxation. In the long run, however, capital is rather mobile internationally. Indeed, capital owners can escape taxes by moving their capital abroad. This implies that the tax-shifting effect to capital is unlikely to occur in the long run. Moreover, with capital supplied elastically, tax shifting towards capital may raise the efficiency cost of taxation, since capital is the overtaxed factor.

2.3.2   Tax shifting across countries

A second extension that allows for a tax-shifting effect is endogenizing the terms of trade (see chapter 5). In particular, if the home country exerts monopsony power on the international market for polluting commodities, tariffs or quotas imposed by the home country would be optimal since they improve national welfare through favorable terms-of-trade effects. However, international agreements may preclude the use of tariffs or quotas. In that case, environmental taxes may operate as implicit trade-policy instruments. Indeed, by reducing the demand for polluting commodities, environmental taxes change the terms of trade in favor of the home country. In this way, the national government can improve the efficiency of the national tax system as a revenue-raising device. Intuitively, the home country shifts the burden of taxation towards the suppliers of polluting commodities. The terms-of-trade improvement for the home country involves a terms-of-trade loss in other countries, however.

2.3.3   Tax shifting between incomes

The third tax-shifting effect appears if we extend the stylized models by including heterogeneous households who receive different types of income, such as labor income, capital income, unemployment benefits, disability benefits, pensions, etc. (see chapter 6). The incidence of environmental taxes may be borne by all these incomes. If the revenues from these environmental taxes are used to reduce taxes that bear only on labor -- e.g. by providing in-work benefits -- the overall tax burden on labor incomes may be reduced. This stimulates the incentives to supply labor and thus alleviates pre-existing distortions on the labor market. However, the lower tax burden on labor is associated with a higher tax burden on non-labor incomes. Political constraints, related to distributional concerns, may prevent governments from shifting the tax burden to non-labor incomes.[16]

The overall conclusion is that, in principle, initial inefficiencies in the tax system may render a strong double dividend feasible because they allow for a tax-shifting effect. However, one may wonder about the origin of the initial inefficiencies. Indeed, it is important to know why governments have not already adopted other (more direct) instruments to alleviate initial inefficiencies, but instead rely on environmental taxes to do so. The next section focuses on the political constraints that may prevent tax-shifting effects.

2.4   Political-economy issues

Coase (1960) has argued that a market economy is able to arrive at the optimal allocation of goods -- even in the presence of environmental externalities. A crucial condition in Coase's theorem is that the property rights of the environment are well defined and that transaction costs are absent. In that case, polluters negotiate with victims about the amount of pollution, and a financial transfer is made to the owner of the property rights. In reality, transaction costs impede direct negotiations between polluters and victims. Therefore, the government usually represents the interests of the victims in negotiations with polluters.

In Coase's theorem, the distribution of property rights is irrelevant for the allocation of goods in equilibrium -- and thus for environmental quality. For the distribution of incomes between polluters and victims, however, the distribution of property rights is important. This makes it relevant also from a political point of view. Indeed, the distribution of property rights in most countries is

---

[16] A similar tax-shifting effect operates in models with an imperfect labor market. In particular, environmental tax reforms may reduce the fall-back position of the union if they favor real after-tax wages compared to income during unemployment. In that case, the reform alleviates the non-tax distortion associated with labor-market imperfections, thereby allowing for a positive second dividend. However, this non-environmental dividend is positive only because the reform redistributes incomes from the unemployed towards the employed (see chapter 7).

determined by a continuous process of negotiations between polluters and governments (representing the victims). This translates into a political discussion about the choice of environmental policy instruments used by the government. In particular, by making a decision about the environmental policy instrument, the government implicitly determines the distribution of property rights. To illustrate, if the government imposes environmental taxes, it nationalizes the property rights which typically were previously in the hands of polluters (who had the right to pollute freely). Polluters will perceive this nationalization of property rights as inequitable, which may cause political tensions. To avoid adverse equity effects, the government may therefore use the revenues from environmental taxes to compensate polluters. However, in general, the government does not have the appropriate instruments or information to neutralize all distributional effects. Accordingly, the distributional consequences are often a complication for the introduction of environmental taxes. Environmental policy instruments that raise no revenues do not suffer from such complications since they do not cause a redistribution of property rights.

Compensating polluters for the loss of property rights may also be inefficient compared to other ways of revenue recycling. Indeed, in a second-best world, the distribution of property rights determines not only who bears the incidence of the social costs of pollution abatement -- i.e. the polluter or the victim -- but also the magnitude of the abatement costs. To illustrate, Parry (1997) and Goulder et al. (1997) suggest that, in the presence of prior tax distortions, the efficiency costs of non-revenue-raising instruments are substantially higher than the efficiency costs of environmental taxes. This is because, unlike non-revenue-raising instruments, environmental policy instruments that raise revenue can reduce other distortionary taxes. In the terminology of Goulder et al., non-revenue-raising instruments cause only a tax-interaction effect, whereas taxes involve also a revenue-recycling effect. Also Fullerton and Metcalf (1996) emphasize the relationship between the distribution of property rights and the social cost of pollution abatement. They argue that environmental policies that do not raise revenues may create scarcity rents that are left in private hands. This may form a barrier for new firms to enter the market, thereby raising the price of output and possibly exacerbating pre-existing distortions. Environmental charges tax away these scarcity rents so that they do not suffer from this additional efficiency cost of environmental policy.[17]

There exists, thus, a basic conflict between political viability and economic efficiency of environmental taxes. If an environmental tax reform aims at not only environmental objectives, but also other goals, this requires that the tax-shifting effect is sufficiently large. Indeed, the government needs to adopt the environmental tax as an instrument to redistribute the burden of taxation. However, political considerations typically call for a minimization of distributional effects. In particular, if governments redistribute incomes under the flag of an environmental tax reform, this will reduce the probability that consensus will be reached about the introduction of environmental taxes. Hence, if governments adopt environmental taxes to obtain a second, non-environmental dividend, they may reduce the possibility that these taxes will indeed be introduced. This is just the opposite of what a number of policy makers had in mind when they used the double dividend as an argument in favor of an environmental tax reform.

Political discussions about environmental policy instruments are often dominated by distributional concerns, rather than efficiency concerns. Therefore, environmental taxes have not been adopted on a large scale in most countries. In countries where they have been implemented, the revenues from environmental taxes seem to be more of a complication than an advantage. To illustrate, the Netherlands has introduced a tax on energy consumption of households in 1996. To avoid changes in the income distribution, the Dutch government introduced a tax-free allowance for unavoidable energy use of households. In this way, the revenues from environmental taxes were minimized and the distributional consequences mitigated. Other Northern European countries have also imposed energy taxes on firms, the revenues of which were recycled through targeted

---

[17] Fullerton and Metcalf (1997) argue that the emphasis of the double-dividend literature on the revenue-raising capacity of pollution taxes is misplaced. In their view, it is crucial whether environmental policies create scarcity rents that are left in private hands, rather than whether environmental policies yield public revenue. However, the choice between revenue-raising versus non-revenue-raising instruments implicitly determines who reaps scarcity rents.

subsidies for the polluting industries. Furthermore, in exchange for emission reductions, energy-intensive industries were exempt from the energy tax.

2.5   Interactions between dividends

In the discussion above, we have implicitly assumed that there are no feedback effects of the environment on the economy. Indeed, the welfare effects of an environmental tax reform have been divided into two separate and independent dividends: an environmental dividend and a non-environmental dividend. If one allows for feedback mechanisms from environmental quality on the behavior of households or on productivity, the environmental dividend and the non-environmental dividend partly coincide. Indeed, changes in environmental quality directly affect non-environmental welfare. Accordingly, it is not straightforward to divide the welfare effects of an environmental tax reform into two independent dividends. It seems natural, however, to attribute the feedback effects of environmental quality on the economy to the non-environmental dividend. In that case, an environmental tax reform may increase the scope for a strong double dividend (e.g. if a better quality of the environment yields favorable effects on private behavior or on productivity of private production factors); (see chapter 8).

Environmental quality may also exert feedback effects on the distribution of welfare across households. In particular, the discussion in section 2.4 assumes that the environmental benefits are equally distributed across households. In that case, the tax-shifting effect may be politically unacceptable, as it involves a redistribution of welfare across individuals. To illustrate, a reform from labor towards pollution taxation that enhances efficiency may not be politically acceptable because the costs of the environmental policy are borne disproportionally by poor households, while the benefits are equally distributed over rich and poor households. However, there may be a justification for such a tax-shifting effect if the environmental benefits are not equally distributed. This has been illustrated by Bovenberg and van Hagen (1997). They allow poor households to feature a relatively high concern for the environment, e.g. because they live in a more polluted area. In that case, the environmental benefits serve as an alternative way to redistribute welfare in favor of poor households. This allows for less redistribution through the tax system. Accordingly, a tax-shifting effect from poor to rich households may become politically acceptable. Indeed, an environmental tax, the revenues of which are recycled through lower labor taxes, may reduce the inefficiencies associated with income redistribution, while still maintaining the distribution of welfare in a broad sense (i.e. including environmental welfare). The result by Bovenberg and van Hagen illustrates that there may be a case for linking the two issues of protecting the environment and alleviating prior tax distortions.

2.6   Conclusions

The double-dividend literature suggests that environmental tax reforms are unlikely to improve the quality of the natural environment and, at the same time, enhance the efficiency of the tax system from a revenue-raising point of view. Indeed, a strong double dividend is possible only if the reform induces a tax-shifting effect, i.e. a shift in the tax burden between factors, countries or incomes. These tax-shifting effects are unlikely to occur because international agreements or political constraints preclude large shifts in the distribution of income. Indeed, political viability of environmental taxes is strengthened if the revenues from these taxes are used to compensate the polluters for their income losses. However, such a compensation scheme is not compatible with the tax-shifting effect. Hence, there is a basic conflict between the strong double dividend and the political viability of environmental taxes.

The failure of the double dividend in the absence of a tax-shifting effect does not mean that environmental taxes should not be imposed. It does imply, however, that these taxes should be justified on the basis of their positive effects on environmental quality. Burdening these taxes with other, non-environmental objectives typically makes them less likely to be introduced.

3 Environmental tax reform in the benchmark model

This chapter develops a general equilibrium model to analyze the welfare effects of an environmental tax reform in a second-best framework. The economy faces two distortions. First, the model allows for environmental externalities. In particular, production and consumption cause pollution, which individual agents do not take into account when making their decisions. However, as an aggregate, pollution harms household utility through environmental amenities. The second distortion enters the model because lump-sum taxes are not available, so that the government requires distortionary labor taxes to finance its public goods.

The analysis reveals that shifting the tax burden away from labor towards the polluting behavior of households or firms alleviates the environmental distortion. However, it typically exacerbates the labor-market distortion. Therefore, there exists a fundamental trade-off between environmental protection and reductions in tax distortions. This has important implications for the level of the optimal second-best pollution tax.

The rest of this chapter is organized as follows. Section 3.1 elaborates on the model. Section 3.2 explores the effects of a shift in the tax burden away from labor towards pollution by households. In particular, it discusses the effects on employment, environmental quality and welfare. Section 3.3 investigates an environmental tax reform from labor towards pollution taxes on firms. The optimal environmental taxes are derived in section 3.4. Section 3.5 presents some numerical simulations with the model. Finally, section 3.6 concludes.

3.1   The benchmark model

This section describes a general equilibrium model for a small open economy. In the rest of the book, we call it the benchmark model. Subsequent chapters extend the benchmark model in several ways. The structure of the benchmark model is outlined in subsection 3.1.1. Subsection 3.1.2 elaborates on the log-linearized version of the model. Finally, subsection 3.1.3 derives a welfare measure that is used to analyze the welfare effects of environmental taxes.

3.1.1   Structure of the model

Firms
Firm behavior is described by a representative firm that supplies a single commodity (Y) on a perfectly competitive market. The firm maximizes profits subject to a neo-classical production function, $F(L,E)$, which exhibits constant returns to scale with respect to labor (L) and an input that causes pollution when used in production (E). Accordingly, this input is referred to as the polluting input and can be thought off as energy. Solving the firm's maximization problem, we find the following first-order conditions:

$$P_Y \frac{\partial Y}{\partial L} = W(1 + T_L) \tag{3.1}$$

$$P_Y \frac{\partial Y}{\partial E} = P_E + T_E \tag{3.2}$$

where P_Y is the market price of output, W is the market-wage rate, T_L is an ad-valorem tax on labor income, P_E is the market price of the polluting input and T_E is a specific tax on polluting inputs. Relations (3.1) and (3.2) represent the implicit demand equations for, respectively, labor and the polluting input. The assumptions of constant returns to scale, perfect competition and profit maximization imply the following non-profit equation:

$$P_Y Y = (1 + T_L) W L + (P_E + T_E) E \tag{3.3}$$

Households

The utility function for the representative household contains five elements. Three private goods enter utility: leisure (V) and two private consumption commodities. One of these commodities pollutes the environment when consumed and is denoted as the polluting consumption commodity (D). The other is referred to as the clean commodity (C). The two consumption commodities enter utility in a weakly separable way from leisure. Hence, changes in the demand for leisure do not impact the composition of consumption over the two consumption commodities. Furthermore, the subutility derived from the two private consumption goods is homothetic.[18] Apart from private goods, two public goods enter utility: public consumption goods (G) and the quality of the natural environment (M). The individual household does not optimize with respect to these public goods but rather takes the quantities of these two goods as given. Public goods are weakly separable from private goods in utility. This assumption ensures that changes in the provision of public goods do not affect private demands.

Households adopt leisure and the two private consumption commodities as instruments to maximize utility $U = u(g[V, q(C,D)], h(G,M))$, subject to the following budget constraint:

$$W L_S = P_C C + (P_D + T_D) D \tag{3.4}$$

In particular, private consumption is constrained by after-tax labor income (WL_S). The time endowment is normalized at unity so that $L_S = 1 - V$, where L_S denotes labor supply. The consumer price for clean commodities is equal to its market price, denoted by P_C, while an environmental tax on the polluting

---

[18] This implies that, in the absence of environmental externalities, uniform taxes on clean and polluting consumption commodities are optimal. In our model, this is equivalent to a tax on labor income.

commodity (T_D), raises the consumer price of this commodity above its market price (P_D). Solving the household maximization problem, we find the following first-order conditions:

$$\frac{\partial u}{\partial V} = \lambda\, W \tag{3.5}$$

$$\frac{\partial u}{\partial C} = \lambda\, P_C \tag{3.6}$$

$$\frac{\partial u}{\partial D} = \lambda\, (P_D + T_D) \tag{3.7}$$

where the Lagrange multiplier λ denotes the marginal utility of income. Expressions (3.5) - (3.7) represent the implicit expressions for labor supply (i.e. the demand for leisure) and the demand for the two private consumption commodities.

Government
The government supplies public goods (G) and relies on three taxes to finance its spending: an ad-valorem tax on labor income (T_L) and two specific environmental taxes. One of these latter taxes applies to the polluting input (T_E), while the other is levied on polluting consumption commodities (T_D). The government does not issue public debt. Hence, its budget is assumed to be balanced, i.e.:

$$P_G G = T_L W L + T_E E + T_D D \tag{3.8}$$

where P_G is the price that the government pays for the public consumption good. Public consumption is exogenous in the model. This chapter examines the effects of a rise in environmental taxes on either firms or households when the revenues are recycled through a reduction in the tax on labor income. Hence, T_L is endogenously determined to ensure that the government meets its revenue requirements ex post.

Environmental quality

The quality of the natural environment in the model depends negatively on the demand for, respectively, polluting inputs and polluting consumption commodities, i.e. $M = m(E,D)$ where m_E, $m_D < 0$. This relationship neglects spacial issues. Furthermore, in our static model, we abstract from lags between the moment of pollution due to economic activity and the environmental damage associated with stock-flow effects. If domestic pollution would cross international borders, the link between domestic production and consumption and the quality of the natural environment in the domestic country would be weakened.

International trade

All commodities (i.e. the two private consumption commodities, domestic output and the public good) as well as the polluting input are tradable. Accordingly, their before-tax prices are determined on world markets and are exogenous to the small open economy.[19] Labor, in contrast, is assumed to be immobile internationally. Accordingly, the wage rate is the only price that is determined domestically. Flexible wage adjustment ensures equilibrium on the labor-market, i.e. $L = L_S$. In accordance with the law of Walras, we find the equilibrium on the balance of payments by combining the non-profit condition, the household budget constraint and the government budget constraint:

$$P_E E + P_D D + P_C C + P_G G = P_y Y \tag{3.9}$$

The balance-of-payments condition represents the budget constraint for the economy as a whole, i.e. the value of the domestic production of tradables constrains overall domestic demand for tradables.

---

[19] An alternative interpretation of fixed prices P_y, P_E, P_C, P_D and P_G is that these goods are produced domestically with constant rates of transformation. This interpretation may also hold for only some goods, e.g. for clean commodities and the public good, whereas the other goods are imported from abroad at fixed world-market prices. To illustrate, fossil fuels are a good example of tradable polluting commodities.

3.1.2   Linearization

To examine the economic consequences of an environmental tax reform analytically, we log-linearize the model around an initial equilibrium. Table 3.1 contains the log-linearized model. Notation is defined at the end of the table. A tilde (~) denotes a relative change, unless indicated otherwise. Prices of tradable commodities and inputs are given from abroad. We assume that these prices do not change as a result of domestic policies. The supply of public goods (G) is fixed exogenously.

Appendix 3A derives the behavioral relations from the optimization of firms and households. Relation (T3.2) in Table 3.1 represents the demand for polluting inputs at a given level of employment. The second term in this equation denotes the substitution effect caused by changes in relative prices. σ_{LE} represents the substitution elasticity between labor and the polluting input.

Expression (T3.5) reveals that a rise in the after-tax real wage rate (W_R) boosts labor supply if the uncompensated wage elasticity ($\eta_{LL} = V(\sigma_V - 1)$) is positive. In particular, the substitution effect (which reduces the demand for leisure) dominates the income effect (which boosts the demand for leisure) if the substitution elasticity between leisure and private consumption (σ_V) exceeds unity. Empirical evidence suggests that the value of the wage elasticity is generally positive, although rather small (see e.g. Hauseman, 1985). Therefore, the rest of this book assumes that the labor-supply curve is indeed upward bending. The quality of the environment does not directly affect private behavior because of the separability assumptions about utility.

Table 3.1: Linearized benchmark model

Firms

Domestic output	$\tilde{Y} = w_E \tilde{E} + w_L \tilde{L}$	(T3.1)
Polluting-input demand	$\tilde{E} = \tilde{L} - \sigma_{LE} [\tilde{T}_E - (\tilde{W} + \tilde{T}_L)]$	(T3.2)

Non-profit condition $\qquad 0 = w_L(\tilde{W} + \tilde{T}_L) + w_E\tilde{T}_E$ $\qquad$ (T3.3)

Households

Household budget
constraint $\qquad w_L(1 - \theta_L)(\tilde{L} + \tilde{W}) = w_C\tilde{C} + w_D(\tilde{D} + \tilde{T}_D)$ $\quad$ (T3.4)

Labor supply $\qquad \tilde{L}_S = \eta_{LL}\tilde{W}_R$ $\qquad$ (T3.5)

Real after-tax wage $\qquad \tilde{W}_R = \tilde{W} - [w_D/(w_D + w_C)]\tilde{T}_D$ $\qquad$ (T3.6)

Private consumption $\qquad \tilde{C} - \tilde{D} = \sigma_{CD}\tilde{T}_D$ $\qquad$ (T3.7)

Government

Government budget $\qquad 0 = w_L\tilde{T}_L + w_E\tilde{T}_E + w_D\tilde{T}_D +$

constraint $\qquad \theta_L w_L(\tilde{W} + \tilde{L}) + \theta_E w_E\tilde{E} + \theta_D w_D\tilde{D}$ $\quad$ (T3.8)

Labor-market equilibrium $\tilde{L}_S = \tilde{L}$ $\qquad$ (T3.9)

Environmental quality $\qquad w_M\tilde{M} = - \gamma_E w_E\tilde{E} - \gamma_D w_D\tilde{D}$ $\qquad$ (T3.10)

Material balance $\qquad \tilde{Y} = w_C\tilde{C} + (1 - \theta_D)w_D\tilde{D} + (1 - \theta_E)w_E\tilde{E}$ $\quad$ (T3.11)

Endogenous $\qquad \tilde{Y}, \tilde{L}, \tilde{E}, \tilde{D}, \tilde{C}, \tilde{L}_S, \tilde{W}, \tilde{W}_R, \tilde{M}, \tilde{T}_L$

Exogenous $\qquad$ *Policy Instruments* $\qquad \tilde{T}_E, \tilde{T}_D$, $\tilde{G} = 0$

$\qquad\qquad$ *Prices* $\qquad\qquad\qquad \tilde{P}_Y = \tilde{P}_E = \tilde{P}_C = \tilde{P}_D = \tilde{P}_G = 0$

Notation

Y = output $\qquad\qquad\qquad\qquad L$ = employment

E = demand for polluting inputs $\qquad C$ = consumption of clean commodities

L_S = labor supply $\qquad\qquad\qquad D$ = consumption of dirty commodities

W = market wage rate $\qquad\qquad W_R$ = real after-tax wage rate

M = environmental quality $\qquad\quad T_L$ = ad-valorem tax on labor income

T_E = tax on polluting inputs T_D = specific tax on polluting commodities
G = public consumption
P_I = market price for commodity $I = Y,C,D,E,G$

Parameters

σ_{LE} = substitution elasticity between labor and the polluting input in production
σ_{CD} = substitution elasticity between clean and polluting consumption in utility
σ_V = substitution elasticity between leisure and private consumption in utility
η_{LL} = $V(\sigma_V - 1)$ = uncompensated labor-supply elasticity
γ_E = $-m_E/(P_E+T_E)$; $\gamma_D = -m_D/(P_D+T_D)$

Taxes

$$\tilde{T}_L = \frac{dT_L}{1+T_L} \quad ; \quad \tilde{T}_E = \frac{dT_E}{P_E+T_E} \quad ; \quad \tilde{T}_D = \frac{dT_D}{P_D+T_D}$$

$$\theta_L = \frac{T_L}{1+T_L} \quad ; \quad \theta_E = \frac{T_E}{P_E+T_E} \quad ; \quad \theta_D = \frac{T_D}{P_D+T_D}$$

Shares

$w_L = (1+T_L)WL/P_YY$ $w_E = (P_E+T_E)E/P_YY$
$w_C = P_CC/P_YY$ $w_D = (P_D+T_D)D/P_YY$
$w_G = P_GG/P_YY$ $w_M = M/P_YY$

Relationships between shares

From (3.3) $w_E + w_L = 1$ (S1)

From (3.4) $(1 - \theta_L)w_L = w_C + w_D$ (S2)

From (3.8) $w_G = \theta_L w_L + \theta_E w_E + \theta_D w_D$ (S3)

From (3.9) $1 = w_C + w_G + (1 - \theta_E)w_E + (1 - \theta_D)w_D$ (S4)

3.1.3   Welfare

We measure the welfare effects of an environmental tax reform by the marginal excess burden (*MEB*), defined as the compensating variation divided by output. It corresponds to the additional transfer that must be provided to households to keep utility, after a policy shock, at its initial level (i.e. the level without the policy shock). A positive excess burden thus implies a loss in welfare. The *MEB* is called 'excess' burden because it corresponds to the loss in welfare over and above the revenues collected by the government. One can interpret the costs due to the excess burden as the 'hidden' costs of financing public expenditures because they are not reflected in tax revenue.

Dividing the MEB into distortions
The *MEB* can be written as the sum of three market distortions (see appendix 3B):

$$MEB = -\theta_L w_L \tilde{L} \quad - \quad [\theta_E - \frac{u_M \gamma_E}{\lambda}] w_E \tilde{E} \ - \ [\theta_D - \frac{u_M \gamma_D}{\lambda}] w_D \tilde{D}$$

$$\qquad\qquad\qquad\qquad\qquad\qquad\qquad\qquad\qquad\qquad\qquad (3.10)$$

$$\underset{\text{distortion}}{\underbrace{\text{labor-market}}} \qquad\qquad\qquad \underset{\text{distortions}}{\underbrace{\text{environmental}}}$$

where u_M denotes the marginal utility of a cleaner environment. The first term on the RHS of (3.10) involves the impact of the reform on the *labor-market distortion*. Employment yields a first-order welfare gain if the pre-existing labor income tax is positive. Intuitively, the labor tax drives a wedge between the marginal social benefits of employment in terms of additional production and the marginal social opportunity costs in terms of foregone leisure. Indeed, additional production due to more employment not only compensates workers for giving up their leisure, but also yields public revenues.

The second and third terms on the RHS of (3.10) denote the *environmental distortions*, which are due to pollution in production and consumption, respectively. The welfare effect of a marginal increase in pollution is given by the difference between two terms: on the one hand, a tax term (θ_E or θ_D), which measures the marginal social benefits of pollution due to the additional tax revenues associated with a wider tax base and, on the other hand, the marginal environmental damage from pollution ($u_M \gamma_E/\lambda$ or $u_M \gamma_D/\lambda$). If we start from an equilibrium without environmental taxes (i.e. $\theta_E = \theta_D = 0$), a reduction in pollution enhances overall welfare because the marginal social benefits of less pollution exceed the marginal social costs.

In the 'first-best' case in which there is no need to finance public spending through distortionary labor taxes ($\theta_L = 0$), changes in employment would not affect welfare. Indeed, the social opportunity costs of additional employment coincide with the social benefits. In that case, expression (3.10) reveals that it would be optimal to fully internalize the environmental distortions, i.e. $\theta_E = u_M \gamma_E/\lambda$ and $\theta_D = u_M \gamma_D/\lambda$. The optimal tax would thus be equal to the Pigovian tax. At the Pigovian level, the loss in welfare on account of more pollution is exactly offset by the welfare gain on account of the broadening of the tax base. Hence, a marginal change in the demand for polluting commodities and polluting inputs does not affect welfare through the channel of the environmental distortions.

In the presence of a distortionary tax on labor ($\theta_L > 0$), a marginal change in pollution may affect welfare also through the channel of the labor-market distortion. In particular, changes in pollution may affect the level of employment. Whereas this leaves welfare unchanged in a first-best setting without labor taxes, it does impact welfare in a second-best world in which the government requires distortionary labor taxes to meet its revenue requirement. The optimal tax thus typically deviates from the Pigovian level. Whether or not a change in the tax mix improves welfare thus depends not only on the effect on the environmental distortion, but also on the response of employment. If an environmental tax reform raises employment at the Pigovian tax level, it is optimal for the government to set the environmental tax above the Pigovian rate. In contrast, if raising the pollution tax above its Pigovian level reduces employment, it would be optimal to set the environmental tax below the Pigovian rate.

Dividing the MEB into dividends

By rearranging terms, the marginal excess burden can alternatively be written as (see appendix 3B):

$$MEB = \underset{\substack{\text{blue} \\ \text{dividend}}}{-\tilde{B}} - \underset{\substack{\text{green} \\ \text{dividend}}}{\frac{u_M}{\lambda} w_M \tilde{M}}$$

(3.11)

where

$$\tilde{B} = \theta_L w_L \tilde{L} + \theta_E w_E \tilde{E} + \theta_D w_D \tilde{D}$$

(3.12)

The three terms on the RHS of (3.12) stand for the effect on the base of, respectively, the labor tax, the environmental tax levied on firms and the environmental tax on household consumption. We call the relative change in B the tax-base effect. The effect on the tax base reflects the consequences of a different tax mix for the efficiency of the tax system as an instrument to raise revenue. In particular, an erosion of the tax base indicates that the tax system becomes a less efficient instrument as a revenue-raising device. Expression (3.11) indicates that the *MEB* can be divided in a tax-base effect, represented by the first term on the RHS of (3.11), and an effect on the quality of the environment, referred to as environmental welfare. These two terms reflect the twofold task of the tax system. On the one hand, the tax system is used as an instrument to meet revenue requirements in the least distortionary way. In the rest of this chapter we refer to an improvement in this non-environmental welfare component as a *blue* dividend. On the other hand, the tax system aims at internalizing environmental externalities. If welfare improves on account of the environmental component, we call this a *green* dividend. If both the blue and the green dividend are positive, we speak about a *strong double dividend*.

By using the non-profit condition (T3.3) and the government budget constraint (T3.8), we can write non-environmental welfare also as:

$$\tilde{B} = (1 - \theta_L) w_L \tilde{W}_R$$

(3.13)

The RHS of (3.13) stands for the effect on after-tax real wage income enjoyed by households. Accordingly, if the tax system becomes less efficient from a non-environmental point of view (i.e. the tax base effect is negative), real private income drops. Intuitively, the erosion of the tax base means that higher tax rates are required to collect the same amount of revenue. Consequently, although it collects the same amount of revenues, the government has to impose higher marginal tax rates. Accordingly, the loss in private income over and above the collected revenues rises. Although households suffer a loss in welfare if real private income falls, they might nevertheless still gain in terms of overall utility if the improvement in environmental welfare (i.e. the green dividend) more than offsets the adverse effect on non-environmental welfare (i.e. the negative blue dividend, see (3.11)).

3.2   Environmental taxes on households

Appendix 3C solves the model of Table 3.1 analytically for the endogenous variables in terms of the two exogenous policy variables: the environmental tax on polluting inputs and the environmental tax on polluting consumption commodities. The labor tax has been assumed to be endogenous. This section interprets the reduced-form equations. In particular, it discusses how the endogenous variables are affected in case of a budgetary neutral shift in the tax system from labor taxes towards environmental taxes on households. The second column of Table 3.2 presents the reduced-form coefficients representing the effects of a rise in T_D. In particular, we are interested in the question of whether or not a strong double dividend can be obtained and whether or not an environmental tax reform alleviates the labor-market distortions. Hence, the reduced forms for employment, pollution and after-tax real private incomes are of particular interest. We discuss the effects on the endogenous variables if the reform starts from specific initial equilibria, namely, equilibria without a pollution tax, with a positive tax at some arbitrary level, or with the pollution tax set at its Pigovian rate.

3.2.1   Starting without initial environmental taxes

The second column of Table 3.2 reveals that the introduction of a small environmental tax on households -- where the revenues are recycled through a reduction in the tax rate on labor -- affects neither employment nor production (see the first and second rows with $\theta_D = 0$). Intuitively, this tax does not directly impact production. Hence, the marginal productivity of labor and, therefore, the before-tax wage, and the demand for labor remain unaffected. From the non-profit condition (T3.3), we find that the associated lower tax on labor income allows for a rise in the market price of labor (i.e. the wage after labor taxes but before consumption taxes):

$$\tilde{W} = -\tilde{T}_L \tag{3.14}$$

However, the wage rate that affects the incentives to supply labor (see expression (T3.5) in Table 3.1) is not the market wage but the real after-tax wage (i.e. the wage after not only labor taxes but also (indirect) consumption taxes). The environmental tax on households drives a wedge between the market wage (W) and the real after-tax wage (W_R). Accordingly, the wedge between the before-tax and real after-tax wages consists of not only the distortionary tax on labor but also the environmental tax on consumption (see (T3.6)). Whether replacing the labor tax by the environmental tax stimulates labor supply by raising the real after-tax wage depends on whether lower taxes on labor more than offset the effect of the higher environmental tax on the overall wedge. In order to find the cut in labor taxes made possible by the higher environmental tax, one needs the budget constraint of the government (where (3.14) is substituted into (T3.8)):

$$\tilde{T}_L + \frac{w_D}{w_C + w_D}\tilde{T}_D = -\frac{\tilde{B}}{w_C + w_D} \tag{3.15}$$

The RHS of (3.15) equals the tax-base effect. This term is zero if employment is unaffected and if there are no initial environmental taxes. According to (3.14) and (3.15), the higher environmental tax exactly offsets the effect of the lower labor tax on the wedge between the before- and after-tax real wage. Indeed, given the constraint of revenue neutrality and the same bases for the taxes on consumption and labor income, replacing the tax on labor income by an indirect (environmental) tax on consumption affects only the composition of the wedge between before- and after-tax wages without affecting the overall magnitude of this tax distortion.

Table 3.2: Reduced-form coefficients of the benchmark model

	$\tilde{T}_D$	$\tilde{T}_E$
$\Delta \tilde{Y}$	$-\eta_{LL}\theta_D w_D \dfrac{w_C}{w_C + w_D}\sigma_{CD}$	$-\left[\Delta + \theta_E \eta_{LL}\right] w_E \dfrac{\sigma_{LE}}{w_L}$
$\Delta \tilde{L}$	$-\eta_{LL}\theta_D w_D \dfrac{w_C}{w_C + w_D}\sigma_{CD}$	$-\eta_{LL}\theta_E w_E \dfrac{\sigma_{LE}}{w_L}$
$\Delta \tilde{E}$	$-\eta_{LL}\theta_D w_D \dfrac{w_C}{w_C + w_D}\sigma_{CD}$	$-\left[\Delta + \theta_E w_E \eta_{LL}\right] \dfrac{\sigma_{LE}}{w_L}$
$\Delta \tilde{D}$	$-\left[\Delta + (1+\eta_{LL})\theta_D w_D\right] \dfrac{w_C}{w_D + w_C}\sigma_{CD}$	$-(1+\eta_{LL})\theta_E w_E \dfrac{\sigma_{LE}}{w_L}$
$\Delta \tilde{C}$	$\left[\Delta - (1+\eta_{LL})\theta_D w_C\right] \dfrac{w_D}{w_D + w_C}\sigma_{CD}$	$-(1+\eta_{LL})\theta_E w_E \dfrac{\sigma_{LE}}{w_L}$
$\Delta \tilde{B}$	$-\theta_D w_D w_C \sigma_{CD}$	$-(1-\theta_L)\theta_E w_E \sigma_{LE}$

$$\Delta = \left[(1-\theta_L)w_L - \theta_D w_D\right] - \eta_{LL}\left[\theta_L w_L + \theta_E w_E + \theta_D w_D\right] > 0$$

3.2.2 Starting from positive environmental taxes

Starting from a situation without any environmental taxes, a small increase in these taxes would not affect employment and would enhance welfare. This result, however, holds only for very small environmental taxes. In order to gain some insight into the effects of large environmental taxes, we must explore the effects of a marginal increase in environmental taxes, starting from an equilibrium in which environmental taxes are positive.

For the case where the initial environmental taxes are positive (i.e. $\theta_D > 0$) the first and second rows of the second column in Table 3.2 show that an increase in the consumption tax on polluting commodities reduces both employment and production if the uncompensated wage elasticity of labor supply is positive (i.e. $\eta_{LL} > 0$). The negative effect on production and employment is due to a reduction in the real after-tax wage and, therefore, in the incentives to supply labor. The negative effect on the real after-tax wage comes about because the lower tax rate on labor income does not fully compensate workers for the adverse effect of the higher environmental tax on their real after-tax wage. This incomplete offset is due to the erosion of the base of the environmental tax. In particular, the higher environmental tax induces households to switch from polluting to non-polluting consumption commodities. If the initial tax rate on the polluting commodities is positive, this behavioral effect erodes the base of the environmental tax and, therefore, produces a negative *tax-base effect*, also called the *tax-burden effect* (i.e. $\tilde{B} < 0$). Expression (3.15) indicates that the replacement of labor taxes by environmental taxes on consumption widens the wedge between the before- and real after-tax wages (i.e. $\tilde{T}_L + [w_D/(w_C + w_D)]\tilde{T}_D > 0$) on account of this negative effect on the tax base. Thus, if it needs to maintain overall tax revenues, the government is unable to reduce the labor tax sufficiently to offset the adverse effect of the higher environmental tax on the real after-tax wage. The resulting lower income from an additional unit of work reduces labor supply, and therefore, employment. The reduction in the real after-tax wage rate implies that the blue dividend always fails if we start from a positive initial pollution tax.

Accordingly, an environmental tax reform from labor towards pollution taxes on households does not yield a strong double dividend.

The magnitude of the adverse employment effect depends (apart from the uncompensated wage elasticity of labor supply) on both the initial environmental tax and the substitution elasticity in consumption between the polluting and non-polluting commodities. These latter parameters determine also the magnitude of the decline in real after-tax wages. A higher initial tax rate strengthens the adverse revenue effects of the erosion of the base of the environmental tax, thereby reducing the room to cut the tax on labor income. This negatively affects after-tax wages and, therefore, harms the incentives to supply labor.

A higher substitution elasticity between polluting and non-polluting commodities raises not only the positive effects on environmental quality, but also the adverse effects on the incentives to supply labor. Intuitively, it strengthens the erosion of the base of the environmental tax, thereby limiting the scope to reduce taxes on labor income. Thus, a fundamental trade-off exists between positive environmental effects and adverse effects on the incentives to supply labor; the more substantial are the beneficial environmental effects of a given tax on polluting consumption commodities, the larger become the adverse effects on the incentives to supply labor.

Another important determinant of the adverse employment effect is the size of the public sector, which is closely related to the determinant, Δ, of the reduced form.[20] In particular, a larger revenue requirement due to a higher level of public spending implies that an environmental tax yields a larger adverse effect on employment. The reason is that the environmental tax becomes a less effective instrument for raising additional revenue if a larger public sector requires higher initial tax rates on labor and the polluting commodities (i.e. θ_L, θ_E, and θ_D are large). Intuitively, how effective higher environmental tax rates are in raising additional public revenues depends on the balance between, on the one hand, larger revenues from a given tax base and, on the other hand, smaller revenues on account of behavioral changes that erode the tax bases of pre-existing taxes.

---

[20] The determinant can be rewritten as: $\Delta = [(1-\theta_L)w_L - \theta_E w_D] - \eta_{LL}w_G$. Accordingly, the larger public expenditures (w_G) are, the smaller is the determinant.

The relative importance of the second effect (i.e. the tax-base effect) on revenues depends on both the size of the initial tax rates and the strength of the behavioral changes. In the case of a higher environmental tax, two behavioral changes shrink the tax base. First, as discussed above, a change in the composition of the consumption basket away from the polluting commodity erodes the base of the environmental tax. The revenue impact of this first behavioral change depends on the initial tax rate on the polluting consumption commodity. The second behavioral effect reducing the tax base is the adverse employment effect. In particular, higher initial tax rates imply that a fall in employment reduces revenues from labor taxes and environmental taxes more substantially. Consequently, in the presence of a larger public sector, which is associated with high initial tax rates, higher environmental taxes are less effective in raising additional public revenue because they harm employment and, therefore, erode the bases of higher pre-existing taxes. This adversely affects revenue and further diminishes the scope for reducing taxes on labor, thereby negatively affecting labor-supply incentives.[21]

3.2.3   Starting from a Pigovian tax

In the 'first-best' case, in which there is no need to finance public spending through distortionary taxation (i.e. $T_L = 0$), the environmental tax levied on households should be set so that it fully internalizes the external effects of pollution. This yields the following optimality condition: $\theta_D = u_M \gamma_D / \lambda$. If the environmental tax is set at this level, a marginal decline in the environmental tax would not affect overall welfare if the government requires only the Pigovian tax to finance its spending, so that the labor-market distortion is absent (i.e. $T_L = 0$).

---

[21] The model is unstable if the adverse revenue effect of an eroding tax base exceeds the positive revenue effect of a higher tax rate on a given tax base. In that case, the government is beyond the top of the Laffer curve of the environmental tax and $\Delta < 0$. Hence, the tax base of the labor tax, $(1-\theta_L)W_L - \theta_D w_D$, which captures the revenue-raising capacity of a (direct or indirect) tax on labor income under the assumption of no behavioral changes, is smaller than the term $\eta_{LL}(\theta_L w_L + \theta_E w_E + \theta_D w_D)$, which represents the effect of behavioral changes in labor supply on the tax revenues from the existing tax system.

The adverse welfare effects associated with a dirtier environment would exactly offset the positive effects on welfare due to an expansion of the tax base. In this 'first-best' case, labor supply and employment would rise as a consequence of the lower environmental tax. However, without an initial (distortionary) direct tax on labor (i.e. $T_L = 0$), higher employment would not raise welfare because the social opportunity costs of additional employment would exactly offset the social benefits.

In the case where financing of public spending requires a distortionary tax on labor, in contrast, overall welfare would rise if the government would marginally reduce the environmental tax below the level that fully internalizes the external effects of pollution (and would, at the same time, raise the tax rate on labor, T_L, to offset the revenue losses). Intuitively, in the first-best optimum, the environmental benefits exactly balance the costs of the eroding tax base of the environmental tax itself. In a 'second-best' case, in which distortionary taxes are required to raise public revenue, however, one should examine not only how the environmental tax impacts overall tax revenue through its effect on its own base, but also how such a tax affects overall revenue by impacting the bases of other taxes, and thereby the capacity of the rest of the tax system to raise revenue. In particular, in the presence of a distortionary tax on labor, the substitution of an environmental tax for a distortionary labor tax erodes the tax base of the labor tax because employment declines. The environmental tax thus reduces the capacity of the tax system to yield public revenues over and above the revenues from the environmental taxes themselves. Hence, reducing the environmental tax below its 'first-best' level and replacing the revenues by labor taxation would raise welfare because it would alleviate the distortions associated with the financing of public goods. Compared to the 'first-best' case, therefore, the optimal environmental tax would fall if a distortionary tax on labor were needed to finance public spending. Indeed, a Pigovian tax, which is set at a level to fully internalize the environmental damage, is suboptimal in the presence of distortionary taxes aimed at raising revenue.

If the environmental tax is set below its 'first-best' optimum (i.e. $\theta_D < u_M \gamma_D / \lambda$), a higher tax yields ambiguous welfare effects because it reduces not only pollution but also employment. On the one hand, a higher environmental tax enhances welfare by reducing pollution and alleviating the environmental distortion. On the other hand, however, the tax harms welfare by reducing employment, thereby worsening the labor-market distortion. Indeed, this is a classic second-best result: reducing one distortion in an economy with remaining distortions does not necessarily yield positive welfare effects.

3.3   Environmental taxes on firms

The third column of Table 3.2 presents the effects of higher pollution taxes on firms of which the revenues are used to cut labor taxes. Again, we discuss the effects on the endogenous variables if the reform starts from specific initial equilibria.

3.3.1   Starting without initial environmental taxes

If we start from an equilibrium without environmental taxes (i.e. $\theta_E = 0$), the reduced-form coefficient in the second row of Table 3.2 reveals that an environmental tax reform from labor towards polluting inputs leaves employment unchanged. The reason is as follows. The production tax reduces the demand for polluting inputs, which contributes to a cut in domestic production. Hence, labor productivity falls if employment remains constant. This, in turn, causes a decline in the before-tax wage. In particular, the non-profit condition (T3.3) yields the following relationship between the tax on the polluting input and the before-tax wage:

$$\tilde{W} + \tilde{T}_L = -\frac{w_E}{w_L}\tilde{T}_E \tag{3.16}$$

Since polluting inputs are perfectly mobile internationally, the entire burden of the environmental tax can be shifted onto labor in the form of lower labor productivity.

In order to determine the impact on labor supply, we need to examine the effect on the real after-tax wage. This effect corresponds to that on the market wage because the environmental tax on households is unchanged. How the market wage is affected becomes clear from (3.16). In particular, the effect on wages depends on the balance between, on the one hand, the drop in the before-tax wage due to a higher pollution tax and, on the other hand, the cut in the labor tax allowed by the additional revenues from the environmental tax on production. The public budget constraint (T3.8) yields the following relationship between the cut in the tax rate on labor and the rise in the environmental tax on production (where we used (3.16)):

$$\tilde{T}_L + \frac{w_E}{w_L}\tilde{T}_E = -\tilde{B} \qquad\qquad (3.17)$$

The tax-base effect on the RHS of (3.17) is zero if employment does not change because the initial environmental taxes are assumed to be zero. Hence, by combining (3.16) and (3.17), one finds that the after-tax wage remains unaffected; given the constraint of revenue neutrality, the adverse effect of the lower before-tax wage on the after-tax wage is exactly offset by the positive effect of lower taxes on labor income. In a small open economy, substituting a labor tax by a tax on polluting inputs thus amounts to substituting an *explicit* by an *implicit* tax on labor. In the absence of initial implicit taxes, the overall tax burden on labor is unaffected.

Starting from an equilibrium without any environmental taxes, the overall welfare effects of small increases in these taxes are positive. Whereas environmental taxes do not affect employment and, therefore, the distortion in the labor market, they do alleviate the environmental distortion by cutting pollution (see 3.10). Alternatively, from (3.11) we find that small environmental taxes benefit the environment without affecting the capacity of the tax system to raise revenue --as the tax-base effect is zero-- and therefore enhance overall welfare.

3.3.2   Starting from positive environmental taxes

If the initial tax on polluting inputs is positive (i.e. $\theta_E > 0$), the second row in the third column of Table 3.2 reveals that this tax harms employment. The negative effect on employment originates in a fall in the real after-tax wage and the associated adverse effects on the incentives to supply labor. The reason for the decline in the marginal income from work is that the pollution tax reduces the demand for polluting inputs. This adversely affects real labor income through two channels. First, it decreases the marginal productivity of labor and, therefore, the before-tax wage. Secondly, the higher environmental tax erodes its own base. If the initial tax rate on the polluting input is positive, the smaller tax base reduces revenue, thereby producing a negative tax-base effect (i.e. $\tilde{B} < 0$). Accordingly, given the constraint of revenue neutrality, the government cannot cut the tax rate on labor income sufficiently to fully offset the adverse effect of the fall in before-tax wages on the after-tax wage. Indeed, expressions (3.16) and (3.17) reveal that the after-tax wage declines if the tax base of the initial tax system shrinks (i.e. $\tilde{B} < 0$). Intuitively, the supply of the polluting input, effectively, is infinitely elastic because the market price of these inputs is fixed on the world market. Hence, the incidence of the pollution tax falls on the only immobile factor of production, labor, in the form of lower labor productivity. An environmental tax reform thus replaces an *explicit* by an *implicit* tax on labor. The explicit labor tax is a more efficient instrument to raise public revenues than the implicit labor tax, as the latter not only distorts the labor market but also reduces the demand for the polluting input, thereby eroding the tax base (if $\theta_E > 0$). In fact, unlike a direct tax on labor income, the input tax distorts the production process -- at least if one abstracts from the environmental benefits.[22]

High initial tax rates exacerbate the adverse revenue effects of the fall in employment. In our model, the initial tax on pollution, θ_E, measures the gap between the social benefits and the non-environmental costs of pollution. Hence, the erosion of the tax base of the pollution tax (i.e. the tax-base effect) measures

---

[22] This result is closely related to that derived by Diamond and Mirrlees (1971), who show that aggregate production efficiency requires zero taxes on intermediate inputs.

the non-environmental (i.e. private) costs associated with a cleaner environment. We call the costs of a cleaner environment the *tax-burden effect*: real wages are reduced and, given a positive uncompensated wage elasticity of labor supply, η_{LL}, employment drops. The overall burden of taxation is thus exacerbated.

The effect on the labor-market distortion is small if firms have little scope to substitute labor for polluting inputs (i.e. σ_{LE} is small). In that case, however, also the environmental benefits are small. A fundamental conflict thus exists between the objectives of improving environmental quality and minimizing the costs of financing public spending. Indeed, a swap of environmental taxes on firms for labor taxes does not yield a strong double dividend.

3.3.3   Starting from a Pigovian tax

If the initial tax structure fully internalizes the environmental distortion (i.e. $\theta_E = u_M \gamma_E / \lambda$), an ecological tax reform does not produce any first-order welfare effects through the channel of the environmental distortion. Hence, welfare effects are determined solely by the effects on the labor-market distortion (see (3.10)). The reduced-form coefficient in the second row of Table 3.1 reveals that, starting from a Pigovian tax (i.e. $\theta_E = u_M \gamma_E / \lambda > 0$), an environmental tax reform reduces employment if the uncompensated wage elasticity of labor supply is positive. With a positive distortionary tax on labor, the drop in employment exacerbates the distortion on the labor market. The relative importance of the effect on the labor-market distortion depends importantly on the overall revenue requirement. Accordingly, a higher level of public spending reduces the welfare gains from switching from labor to environmental taxation. Indeed, such a change in the tax mix may harm welfare, even though the initial environmental tax is below the level that fully internalizes the external costs of pollution. Reducing the pollution tax further below the Pigovian level may raise welfare by alleviating the labor-market distortion.

3.4   Optimal environmental taxes

Sections 3.2 and 3.3 reveals that the optimal second-best environmental tax is strictly positive but smaller than the Pigovian tax. This section derives expressions for the optimal environmental taxes on households and firms. To that end, we substitute the reduced-form equations for wages and pollution from Table 3.2 into the *MEB* in (3.11) and set it equal to zero. Thus, we find for the optimal pollution taxes on firms and households:[23]

$$\theta_E = \frac{1}{\eta} \frac{u_M \gamma_E}{\lambda} \quad ; \quad \theta_D = \frac{1}{\eta} \frac{u_M \gamma_D}{\lambda} \tag{3.18}$$

$$\eta = \frac{1}{1 - T_L \eta_{LL}} \tag{3.19}$$

where η denotes the Marginal Cost of Public Funds (MCPF). It measures how scarce public funds are relative to private funds. If the government has access to lump-sum taxes, public funds are no scarcer than private income and the MCPF equals unity (see (3.19) with $T_L = 0$). In that case, the optimal pollution tax is equal to the Pigovian tax (see (3.18) with $\eta = 1$). However, if lump-sum taxes are absent and Pigovian taxes do not yield sufficient revenues to meet public revenue requirements, the government requires distortionary taxes to finance its spending. In that case, the MCPF typically lies above unity, as taxation distorts economic decisions. Hence, public funds are scarcer than private funds.[24] In that case, the expressions in (3.18) show that the optimal environmental tax is smaller than the Pigovian rate. This is consistent with our results from sections 3.2 and

---

[23] In deriving the optimal pollution tax on firms (households), we abstract from environmental distortions associated with the polluting behavior of households (firms). Hence, interactions between the two environmental distortions are ignored. The optimal tax rates in (3.18) can alternatively be derived from setting up an optimal tax problem for the government, see e.g. Bovenberg and van der Ploeg (1994a).

[24] If labor is supplied inelastically (i.e. $\eta_{LL} = 0$), neither implicit nor explicit taxes distort the labor supply decision, so that the MCPF equals unity. If the uncompensated wage elasticity of labor supply is negative, the MCPF is smaller than unity.

3.3 which suggest that, if we start from the Pigovian level, a marginal reduction of pollution taxes increases welfare. Intuitively, environmental quality amounts to a public consumption good. A high MCPF makes the supply of this public good more expensive. Hence, it calls for a pollution tax below its first-best level.

These results are consistent with the optimal-tax result derived by Sandmo (1975). That study analyzes the case of a government employing a commodity tax system to simultaneously achieve two objectives: first, to satisfy its revenue requirement and, second, to internalize external effects. It derives the optimal tax rate for an externality-creating commodity as the weighted average of two terms (see also expression (2.1) in chapter 2). First is a Ramsey tax term, which is equal to the inverse elasticity formula familiar from the theory of optimal taxation. Second is a Pigovian tax component, which is related to the marginal social damage of the polluting commodity. In Sandmo's formula, the weight of the Pigovian component decreases as the revenue requirement increases (i.e. if the MCPF becomes larger). Intuitively, an environmental tax reduces pollution by inducing taxpayers to avoid taxes. Tax avoidance not only reduces pollution, however, but also requires the government to levy higher distortionary taxes to finance a given level of public spending. Accordingly, the larger the government's revenue needs are (and hence the higher the distortionary taxes need to be), the less the government can afford tax differentiation aimed at environmental protection. Indeed, marginal tax revenue becomes more valuable if the government has to rely on distortionary taxes to finance its spending. The optimal environmental tax balances the social costs of pollution against the social benefits from additional tax revenues. Therefore, the higher the social value attached to tax revenue, the higher the marginal social costs of pollution have to be to justify environmental taxes. There exists thus a basic conflict between using the tax system to raise revenue and using it to improve environmental quality.

Discussion

The optimal tax formula in (3.18) is consistent with Sandmo's formula because, in our model, the Ramsey component is zero. In that case, the optimal pollution tax is lower than the Pigovian rate if the MCPF exceeds one. This result has been criticized by Fullerton (1997) and Schöb (1997), who argue that the Ramsey component is zero only for a specific normalization of the tax structure. In particular, by using a model similar to that presented in this chapter, they optimize the tax system with respect to taxes on clean commodities, dirty commodities and labor. One of these tax rates should be normalized to zero. For the equilibrium outcome, it does not matter which tax. However, the normalization does matter for the Ramsey component and, therefore, for the magnitude of the optimal pollution tax.

This criticism of Fullerton and Schöb can be understood from section 3.2. In particular, by exploring the replacement of labor taxes by taxes on polluting commodities, we have implicitly selected the clean consumption commodity as the untaxed good. In that case, the government finds it optimal from a non-environmental point of view to adopt only labor taxes to raise public revenues and to set the pollution tax equal to zero. In other words, the Ramsey component is zero. If, in contrast, the labor tax were chosen as the untaxed good, the optimal commodity tax structure in the absence of environmental externalities would not differentiate between polluting and non-polluting commodities. This is because our utility structure implies, on the one hand, that the substitution elasticities between the untaxed commodity (leisure) and the two consumption commodities are equal and, on the other hand, that the subutility function of the two private commodities is homothetic. Hence, the Ramsey term for the optimal pollution tax would be positive. In the presence of environmental externalities, this positive Ramsey term may raise the optimal pollution tax above its Pigovian rate.

In their reply to Fullerton, Bovenberg and de Mooij (1997a) maintain that the clean consumption good is the most appropriate choice for the untaxed good for two reasons. First, the normalization is the most appropriate in the context of the double-dividend debate. Indeed, it clearly isolates the focus on the double dividend by emphasizing tax differentiation between clean and dirty commodities. When people talk about pollution taxes in the context of the double dividend, they typically have this tax differentiation in mind. To illustrate, a carbon tax is generally referred to as the additional tax on carbon compared to that on other commodities. When we talk about a carbon tax of $50 per ton, we do not include the labor tax that bears on carbon (and other commodities). In practice, if only for administrative reasons, consumption taxes are rather flat. Environmental taxes are the taxes that are imposed on top of these consumption (and labor) taxes.

The second reason we selected the clean commodity as the untaxed good is that it clearly reveals how the intuition of the double-dividend argument goes wrong. This argument implies that the government should resort to higher environmental taxes when revenue needs increase. By selecting the clean good as the untaxed good, we show that something is wrong with this intuition. In particular, an increase in the MCPF decreases, rather than increases, the Pigovian component. Hence, higher revenue needs should induce the government to decrease the pollution tax if the Ramsey component is zero. In case of a different normalization rule, this argument might get lost in a discussion about the impact of the change in the MCPF on the weight of the Ramsey component.

3.5   Numerical simulations

This section illustrates our analytical results numerically by presenting some simulations with the benchmark model.

3.5.1   Calibration

The calibration of the model requires empirical information about various shares and elasticities. The shares are based on a rough approximation of most West European economies (see the first and second column in Table 3.3). For the polluting input and the polluting consumption commodity, we take energy data. EU economies are characterized by a large share of labor in production compared to energy. Furthermore, initial labor taxes are large as compared to initial energy taxes. Indeed, T_L is set at 1, which implies a labor tax on a before-tax basis of 50%. The energy tax on firms is set equal to 0.1. Households spend three-quarters of their income on clean commodities and one quarter on energy consumption. The tax rate on energy consumed by households is twice as large as the corresponding tax rate on firms.[25] The share of government spending is quite substantial, as is consistent with most EU economies.

The third column of Table 3.3 presents the values for the elasticities. The substitution elasticity between labor and the polluting input is set equal to 0.4.[26] Empirical evidence on labor supply elasticities suggest an average value of 0.2. Substitution between energy and other consumption commodities is set at 0.5.

Table 3.3: Calibration of shares and parameters of the benchmark model

Shares	Taxes	Elasticities
$w_L = 0.8$	$\theta_L = 0.5$	$\sigma_{LE} = 0.4$
$w_E = 0.2$	$\theta_E = 0.1$	$\sigma_{CD} = 0.5$
$w_D = 0.1$	$\theta_D = 0.2$	$\sigma_V = 3.0$
$w_C = 0.3$		$V = 0.1$
$w_G = 0.44$		$\eta_{LL} = 0.2$

3.5.2   Simulation results

Table 3.4 shows the effects of an increase in energy taxes that raises the producer or consumer price of energy by 10%.[27] It suggests that the energy tax on firms reduces energy consumption by firms by more than 5% in the benchmark model. This green dividend is accompanied by a drop in private income of 0.14% and a reduction in employment of 0.07%. Energy taxes on households imply a 4%

---

[25] In Europe, households typically face higher energy tax rates than firms; see Ekins and Speck (1997).

[26] Chapter 4 discusses empirical evidence on the substitution elasticity between labor and energy.

[27] Our linearized model is suitable only to explore the consequences of marginal policy changes. Although an increase of 10% is somewhat large, it should be stressed that our simulations serve only illustrative purposes.

reduction in energy consumption. At the same time, employment falls by 0.05% and private income drops by 0.1%.

Table 3.4: Effects (in % changes) of an increase in energy taxes that raises the producer/consumer price of energy by 10% according to the benchmark model

	10% tax on firms	10% tax on households
Employment	−0.07	−0.05
Energy use by firms	−5.07	−0.05
Energy use by households	−0.41	−4.06
Private income	−0.14	−0.10

3.6   Conclusions

This chapter shows that pollution taxes exacerbate preexisting tax distortions on the labor market. Therefore, the optimal second-best pollution tax is below the Pigovian level. The reason is that the environment is a collective good; all residents benefit -- irrespective of the amount of labor they supply. Indeed, an improvement in the environment can be interpreted as an increase in the supply of collective goods. All public goods, including the cleaner environment, are ultimately paid for by the immobile production factor, labor. Hence, the abatement costs associated with a cleaner environment reduce the incentives to supply labor at the margin, thereby contracting the base of the labor tax and raising the marginal efficiency costs of financing (other) public spending.

This chapter also reveals that the strong double dividend typically fails. Indeed, although an environmental tax reform yields environmental benefits (i.e. a positive first dividend), it fails to improve the efficiency of the tax system as a revenue-raising instrument (i.e. it yields a negative second dividend).

Appendix 3A   Behavioral relations

This appendix derives the linearized factor-demand relations and expressions for the demand for consumption and leisure.

Factor demand relations

$Y = F(E,L)$ is homogenous of degree one in E and L. Hence, $Y/L = F(E/L,1) \rightarrow y = f(e)$, where $y = Y/L$ and $e = E/L$. The first-order conditions from the firm maximization problem are:

$$\frac{W(1+T_L)}{P_Y} = \frac{\partial Y}{\partial L} = f - ef' \tag{3A.1}$$

$$\frac{P_E + T_E}{P_Y} = \frac{\partial Y}{\partial E} = f' \tag{3A.2}$$

The substitution elasticity between E and L is defined as:

$$\frac{1}{\sigma_{LE}} = -\frac{\partial(\frac{\partial Y/\partial E}{\partial Y/\partial L})}{\partial(\frac{E}{L})} \frac{\frac{E}{L}}{\frac{\partial Y/\partial E}{\partial Y/\partial L}} = -\frac{ef''}{f'} \frac{f}{f-ef'} \tag{3A.3}$$

Linearizing the first-order conditions in (3A.1) and (3A.2), we find (using $\tilde{P}_Y = \tilde{P}_E = 0$):

$$\tilde{W} + \tilde{T}_L - \tilde{T}_E = -\frac{ef''}{f'} \frac{f}{f-ef'} \tilde{e} = \frac{1}{\sigma_{LE}} \tilde{e} \tag{3A.4}$$

Expression (3A.4) is equivalent with (T3.2).

Consumption and labor supply

Dividing (3.6) and (3.7), we get the following expression:

$$\frac{P_D + T_D}{P_C} = \frac{\partial u / \partial D}{\partial u / \partial C} \tag{3A.5}$$

Define:

$$\frac{1}{\sigma_{CD}} = - \frac{\partial(\frac{\partial u / \partial D}{\partial u / \partial C})}{\partial(\frac{D}{C})} \frac{\frac{D}{C}}{\frac{\partial u / \partial D}{\partial u / \partial C}} \qquad (3A.6)$$

Linearizing (3A.5), we find (using $\tilde{P}_C = \tilde{P}_D = 0$):

$$\tilde{T}_D = \frac{\partial(\frac{\partial u / \partial D}{\partial u / \partial C})}{\frac{\partial u / \partial D}{\partial u / \partial C}} = - \frac{1}{\sigma_{CD}} (\tilde{D} - \tilde{C}) \qquad (3A.7)$$

where we used (3A.6). Expression (3A.7) is equivalent with (T3.7).

To derive the expression for labor supply, define $Q = q(C,D)$ as the composite consumption bundle with an ideal price P_Q. This price is a weighted average of P_C and $(P_D + T_D)$. From the household maximization problem, we have:

$$\frac{P_Q}{W} = \frac{\partial u / \partial Q}{\partial u / \partial V} \qquad (A3.8)$$

Define:

$$\frac{1}{\sigma_V} = - \frac{\partial(\frac{\partial u / \partial Q}{\partial u / \partial V})}{\partial(\frac{Q}{V})} \frac{\frac{Q}{V}}{\frac{\partial u / \partial Q}{\partial u / \partial V}} \qquad (3A.9)$$

Linearizing (3A.8), we arrive at:

$$\tilde{P}_Q - \tilde{W} = \frac{\partial(\frac{\partial u / \partial Q}{\partial u / \partial V})}{\frac{\partial u / \partial Q}{\partial u / \partial V}} = - \frac{1}{\sigma_V} (\tilde{Q} - \tilde{V}) \qquad (3A.10)$$

From the definitions of Q and P_Q, we have (with $\tilde{P}_C = \tilde{P}_D = 0$):

$$\tilde{Q} = \frac{w_C}{w_C + w_D} \tilde{C} + \frac{w_D}{w_C + w_D} \tilde{D} \qquad (3A.11)$$

$$\tilde{P}_Q = \frac{w_D}{w_C + w_D} \tilde{T}_D \tag{3A.12}$$

Using the household budget constraint (T3.4), $\tilde{Q}$ can be rewritten as:

$$\tilde{Q} = \tilde{L} + \tilde{W}_R \tag{3A.13}$$

Substituting (3A.12) and (3A.13) into (3A.10), to eliminate $\tilde{Q}$ and $\tilde{P}_Q$, we arrive at:

$$\tilde{V} - \tilde{L} = (1 - \sigma_V) \tilde{W}_R \tag{3A.14}$$

From the definition of V, we have:

$$\tilde{V} = -\frac{L}{V} \tilde{L} \tag{3A.15}$$

Combining (3A.14) with (3A.15), we find (T3.5).

Appendix 3B   Marginal excess burden

In analogy of Keller (1980), we compute the marginal excess burden (*MEB*) as the compensating variation, defined as the transfer necessary to maintain, after a policy change, utility at its original level. For small changes, the compensating variation equals the equivalent variation.

We find the compensating transfer such that, after the policy shock, utility is unchanged compared to the initial equilibrium, i.e.

$$0 = dU = \frac{\partial u}{\partial V} dV + \frac{\partial u}{\partial C} dC + \frac{\partial u}{\partial D} dD + \frac{\partial u}{\partial M} dM \tag{3B.1}$$

where we used $dG = 0$. Utility maximization implies that the partial derivatives on the RHS of (3B.1) are equal to λ, the Lagrange multiplier corresponding to the budget constraint, times the corresponding price (see (3.5) - (3.7)). Hence, we write (3B.1) as:

$$-\lambda W dL + \lambda P_C dC + \lambda (P_D + T_D) dD + \frac{\partial u}{\partial M} dM = 0 \qquad (3B.2)$$

where we have used $dL = -dV$. Differentiating the household budget constraint (3.4), we find:

$$L dW + W dL + CV =$$

$$C dP_C + P_C dC + D d(P_D + T_D) + (P_D + T_D) dD \qquad (3B.3)$$

where CV is the compensating variation. Substituting (3B.2) into (3B.3), we find:

$$CV = - L dW + C dP_C + D d(P_D + T_D) - \frac{\partial u}{\partial M} \frac{dM}{\lambda} \qquad (3B.4)$$

We define the MEB as CV relative to the value of production $P_Y Y$. This yields the following expression:

$$MEB = - (1 - \theta_L) w_L \tilde{W} + w_D \tilde{T}_D - \frac{u_M w_M}{\lambda} \tilde{M} \qquad (3B.5)$$

where we have used $dP_C = dP_D = 0$. Alternatively, using (T3.10,) we can write the MEB as:

$$MEB = - (1 - \theta_L) w_L \tilde{W}_R + \frac{u_M}{\lambda} (\gamma_D w_D \tilde{D} + \gamma_E w_E \tilde{E}) \qquad (3B.6)$$

In order to derive the expression for the excess burden in (3.10) and (3.11), we use the public budget constraint (T3.8):

$$\theta_L w_L \tilde{L} + \theta_E w_E \tilde{E} + \theta_D w_D \tilde{D} =$$

$$- \theta_L w_L \tilde{W} - w_E \tilde{T}_E - w_D \tilde{T}_D - w_L \tilde{T}_L \qquad (3B.7)$$

Substituting (T3.3) into (3B.7) to eliminate $\tilde{T}_L$, we find:

$$(1 - \theta_L) w_L \tilde{W}_R = \theta_L w_L \tilde{L} + \theta_E w_E \tilde{E} + \theta_D w_D \tilde{D} \qquad (3B.8)$$

Accordingly, by combining (3B.6) and (3B.8), one verifies that (3.10) and (3.11) hold.

Appendix 3C   Analytical solution of the model

This appendix derives analytical expressions for the reduced-form equations presented in Table 3.2. Substituting (T3.11) into (T3.1), one can eliminate $\tilde{Y}$:

$$w_L \tilde{L} = - \theta_E w_E \tilde{E} + (1 - \theta_D) w_D \tilde{D} + w_C \tilde{C} \qquad (3C.1)$$

We express the three endogenous variables on the RHS of equation (3C.1) in terms of $\tilde{L}$ and $\tilde{W}_R$. To find an expression for $\tilde{E}$, we substitute (T3.3) into (T3.2) to eliminate $\tilde{W} + \tilde{T}_L$:

$$\tilde{E} = \tilde{L} - \frac{\sigma_{LE}}{w_L} \tilde{T}_E \qquad (3C.2)$$

To find $\tilde{C}$ and $\tilde{D}$ in terms of $\tilde{W}_R$ and $\tilde{L}$, we combine (T3.4), (T3.6) and (T3.7) to get:

$$\tilde{C} = \tilde{W}_R + \tilde{L} + \frac{w_D}{w_D + w_C} \sigma_{CD} \tilde{T}_D \qquad (3C.3)$$

$$\tilde{D} = \tilde{W}_R + \tilde{L} - \frac{w_C}{w_D + w_C} \sigma_{CD} \tilde{T}_D \qquad (3C.4)$$

Substituting (3C.2), (3C.3), and (3C.4) into (3C.1) one derives:

$$[\theta_L w_L + \theta_E w_E + \theta_D w_D] \tilde{L} = [(1 - \theta_L) w_L - \theta_D w_D] \tilde{W}_R$$

$$+ \theta_E w_E \frac{\sigma_{LE}}{w_L} \tilde{T}_E + \theta_D \frac{w_D}{w_D + w_C} w_C \sigma_{CD} \tilde{T}_D \qquad (3C.5)$$

where we used (S2).

Together with labor supply in (T3.5), (3C.5) can be written in matrix notation as:

$$
\begin{pmatrix} (w_L\theta_L + w_E\theta_E + w_D\theta_D) & -[(1-\theta_L)w_L - \theta_D w_D] \\ 1 & -\eta_{LL} \end{pmatrix} \begin{pmatrix} \tilde{L} \\ \tilde{W}_R \end{pmatrix}
$$

$$
= \begin{pmatrix} \theta_D \dfrac{w_D}{w_C + w_D} w_C \sigma_{CD} & \theta_E w_E \dfrac{\sigma_{LE}}{w_L} \\ 0 & 0 \end{pmatrix} \begin{pmatrix} \tilde{T}_D \\ \tilde{T}_E \end{pmatrix}
\tag{3C.6}
$$

We can solve for the endogenous variables by inverting the matrix on the left-hand side of (3C.6):

$$
\Delta \begin{pmatrix} \tilde{L} \\ \tilde{W}_R \end{pmatrix} = \begin{pmatrix} -\eta_{LL} & [(1-\theta_L)w_L - \theta_D w_D] \\ -1 & (w_L\theta_L + w_E\theta_E + w_D\theta_D) \end{pmatrix}
$$

$$
\begin{pmatrix} \theta_D \dfrac{w_D}{w_C + w_D} w_C \sigma_{CD} & \theta_E w_E \dfrac{\sigma_{LE}}{w_L} \\ 0 & 0 \end{pmatrix} \begin{pmatrix} \tilde{T}_D \\ \tilde{T}_E \end{pmatrix}
\tag{3C.7}
$$

where $\Delta = [(1-\theta_L)w_L - \theta_D w_D] - \eta_{LL}(w_L\theta_L + w_E\theta_E + w_D\theta_D)$ must be positive for the equilibrium to be stable. The reduced-form equations for $\tilde{W}_R$ and $\tilde{L}$ can now be computed from (3C.7):

$$
\Delta \tilde{L} = -\eta_{LL}\theta_D \dfrac{w_C}{w_D + w_D} w_C \sigma_{CD} \tilde{T}_D - \eta_{LL}\theta_E w_E \dfrac{\sigma_{LE}}{w_L} \tilde{T}_E
\tag{3C.8}
$$

$$
\Delta \tilde{W}_R = -\theta_D \dfrac{w_D}{w_C + w_D} w_C \sigma_{CD} \tilde{T}_D - \theta_E w_E \dfrac{\sigma_{LE}}{w_L} \tilde{T}_E
\tag{3C.9}
$$

Using these expressions, the reduced forms for consumption, polluting inputs and output can be derived from, respectively (3C.3) and (3C.4), (3C.2) and (T3.1).

4   The role of capital mobility and factor substitution

This chapter extends the previous analysis by exploring how inefficiencies in the initial tax system affect the potential for a double dividend. In particular, by modeling two non-polluting production factors (i.e. labor and capital), rather than just one, we are able to allow for an inefficient distribution of the tax burden over these two factors. Within such a setting, an environmental tax reform affects private welfare through two channels. First, it induces a tax-burden effect that is responsible for the failure of the double-dividend hypothesis in the previous chapter. With two clean production factors, however, an environmental tax reform affects private welfare through a second channel, namely the distribution of the tax burden over the two production factors. In particular, by redistributing the tax burden across the two factors, the reform affects the efficiency of the tax system as a revenue-raising device if the initial distribution of the tax burden is inefficient from a non-environmental point of view. This effect is called the *tax-shifting effect*. If the reform shifts the tax burden from the overtaxed factor (i.e. the efficient factor) towards the undertaxed factor (i.e. the inefficient factor), the tax-shifting effect alleviates initial inefficiencies in the tax system. However, it exacerbates these inefficiencies if the tax burden is moved onto the factor that is overtaxed already in the initial equilibrium. Hence, initial inefficiencies in the tax system provide both opportunities and dangers for an environmental tax reform. In particular, a double dividend is feasible if, by shifting the tax burden towards the undertaxed factor, the tax-shifting effect makes the tax system more efficient from a non-environmental point of view and is large enough to offset the tax-burden effect.

In illustrating these ideas, we allow for two extreme assumptions about the supply of capital. In the first model with two clean production factors, capital is perfectly mobile internationally. Hence, capital supply, effectively, is infinitely elastic. In that case, raising polluting taxes to cut initial taxes on capital tends to reduce the tax burden in the capital market. Hence, the tax-shifting effect works in the right direction by improving the efficiency of the tax system as a revenue-raising device. Accordingly, a double dividend is obtained if the tax-shifting effect dominates the tax-burden effect.

In the second model, capital is immobile internationally. Moreover, the supply of capital is completely inelastic. Within this framework, we find that substituting environmental taxes for labor taxes may yield a double dividend. Intuitively, by shifting the tax burden from labor (the overtaxed factor) towards the fixed factor (the undertaxed factor), the government may improve the efficiency of the tax system as a revenue-raising device. The second model is similar to that of Bovenberg and van der Ploeg (1996 and 1998b), who also allow for a fixed factor in production. However, in contrast to Bovenberg and van der Ploeg, who allow for involuntary unemployment due to rigid consumer wages, this chapter allows flexible wages to clear the labor market.

The rest of this chapter is structured as follows. Section 4.1 discusses the structure of the model with mobile capital. Section 4.2 analyzes the optimal tax structure of this model, whereas section 4.3 explores the consequences of substituting environmental taxes for taxes on either labor or capital. Section 4.4 discusses the model in which the supply of capital is fixed. Subsequently, the optimal tax structure is presented in section 4.5 and tax reforms are discussed in section 4.6. Section 4.7 presents some numerical illustrations with the models of this chapter. Finally, section 4.8 concludes.

4.1   The model with mobile capital in production

This section extends the benchmark model by including capital in the production function. In order to focus on this extension in the production function, this chapter does not incorporate polluting consumption and pollution taxes on households.

4.1.1   Structure of the model

Three inputs enter the following neo-classical production function $F(L,E,K)$, where K refers to clean capital. The production function exhibits constant returns to scale with respect to the three inputs. Similar to the polluting input, clean capital is mobile internationally, so that its market price is fixed on the world market, i.e. clean capital is in effect supplied infinitely elastic. We assume that all capital is owned by foreigners. Hence, domestic households do not receive capital income.[28] Solving the firm maximization problem, we find the implicit demand functions for labor and the polluting input (see (3.1) and (3.2)) and the following implicit demand for clean capital:

$$P_Y \frac{\partial Y}{\partial K} = R(1 + T_K) \tag{4.1}$$

where R is the before-tax return to capital and T_K is a source-based ad-valorem capital tax. In the presence of capital in production, the non-profit condition looks as follows:

$$P_Y Y = (1 + T_L)WL + (P_E + T_E)E + (1 + T_K)RK \tag{4.2}$$

Compared to the benchmark model, the government has access to an additional tax to raise public revenues, namely, a tax on capital (T_K). The government budget thus amounts to:[29]

$$P_G G = T_L WL + T_E E + T_K RK \tag{4.3}$$

---

[28] Bovenberg and de Mooij (1996) allow for domestic ownership of clean capital. Their analysis shows that domestic ownership does not change the main results as long as the before-tax return to capital is fixed.

[29] Note that D and T_D are absent in this chapter.

The final difference with the benchmark model concerns the material-balance condition, which reads:

$$P_E E + P_C C + P_G G + R K = P_Y Y \tag{4.4}$$

4.1.2   Linearization

Table 4.1 contains the linearized model as far as it differs from the benchmark model. Relations (T4.2) and (T4.3) represent the factor-demand equations for polluting inputs and capital, both relative to employment. The elasticities, ϵ_{ij}^*, (i,j = L,K,E) represent price elasticities of factor demand, conditional on the level of employment. To obtain a better understanding of what determines the own- and cross-price elasticities in these factor-demand equations, Table 4.2 presents these elasticities for three separable production functions, i.e. the polluting input separable, labor separable and capital separable. It reveals that, in all cases, both a higher price of pollution (capital) and a lower price of labor reduce the pollution/labor ratio (capital/labor ratio) (i.e. ϵ_{EE}^* and $\epsilon_{KK}^* > 0$; ϵ_{EL}^* and $\epsilon_{KL}^* < 0$). The signs of the cross elasticities, ϵ_{KE}^* and ϵ_{EK}^* are ambiguous. In particular, a higher price of pollution boosts the demand for capital relative to that of labor if, compared to labor, capital is a better substitute for pollution. Likewise, the pollution/labor ratio expands on account of a higher price of capital if capital is a better substitute for pollution than it is for labor. If inputs are equally good substitutes for each other, the cross-price elasticities are zero.

Table 4.1: Linearized model with mobile capital

Firms

Domestic output	$\tilde{Y} = w_E \tilde{E} + w_L \tilde{L} + w_K \tilde{K}$	(T4.1)
Polluting input demand	$\tilde{E} = \tilde{L} - \epsilon_{EL}^* (\tilde{W} + \tilde{T}_L) - \epsilon_{EE}^* \tilde{T}_E - \epsilon_{EK}^* \tilde{T}_K$	(T4.2)
Capital demand	$\tilde{K} = \tilde{L} - \epsilon_{KL}^* (\tilde{W} + \tilde{T}_L) - \epsilon_{KE}^* \tilde{T}_E - \epsilon_{KK}^* \tilde{T}_K$	(T4.3)

Non-profit condition $\qquad 0 = w_L(\tilde{W} + \tilde{T}_L) + w_E\tilde{T}_E + w_K\tilde{T}_K$ $\qquad$ (T4.4)

Households

Household budget $\qquad w_L(1 - \theta_L)(\tilde{L} + \tilde{W}) = w_C\tilde{C}$ $\qquad$ (T4.5)

Labor supply $\qquad \tilde{L}_S = \eta_{LL}\tilde{W}$ $\qquad$ (T4.6)

Government

Government budget $\qquad \begin{aligned} 0 = {}& w_L\tilde{T}_L + w_E\tilde{T}_E + w_K\tilde{T}_K + \\ & \theta_L w_L(\tilde{W} + \tilde{L}) + \theta_E w_E\tilde{E} + \theta_K w_K\tilde{K} \end{aligned}$ $\qquad$ (T4.7)

Labor-market equilibrium $\tilde{L}_S = \tilde{L}$ $\qquad$ (T4.8)

Environmental quality $\qquad w_M\tilde{M} = -\gamma_E w_E\tilde{E}$ $\qquad$ (T4.9)

Material balance $\qquad \tilde{Y} = w_C\tilde{C} + (1 - \theta_K)w_K\tilde{K} + (1 - \theta_E)w_E\tilde{E}$ $\qquad$ (T4.10)

Notation

E = demand for polluting inputs $\qquad\qquad$ Y = output

L = employment $\qquad\qquad$ K = demand for capital

L_S = labor supply $\qquad\qquad$ W = market wage rate

T_L = ad-valorem tax on labor income $\qquad$ T_K = tax on capital

T_E = tax on polluting inputs $\qquad\qquad$ M = environmental quality

C = consumption of clean commodities

Parameters

ϵ_{IJ}^{*} = demand elasticity (conditional on employment) of factor I with respect to changes in the price of factor J, for $I,J = K,E,L$

η_{LL} = uncompensated labor-supply elasticity

γ_E = $-m_E/(P_E + T_E)$

Taxes

$$\tilde{T}_L = \frac{dT_L}{1+T_L} \quad ; \quad \tilde{T}_E = \frac{dT_E}{P_E+T_E} \quad ; \quad \tilde{T}_K = \frac{dT_K}{1+T_K}$$

$$\theta_L = \frac{T_L}{1+T_L} \quad ; \quad \theta_E = \frac{T_E}{P_E+T_E} \quad ; \quad \theta_K = \frac{T_K}{1+T_K}$$

Shares

$$w_L = (1+T_L)WL/P_YY \qquad w_E = (P_E+T_E)E/P_YY \qquad w_K = (1+T_K)K/Y$$
$$w_C = P_CC/P_YY \qquad\qquad w_G = P_GG/P_YY \qquad\qquad w_M = M/P_YY$$

Relationships between shares

$$w_E + w_L + w_K = 1 \qquad\qquad (1 - \theta_L)w_L = w_C$$

$$w_G = \theta_L w_L + \theta_E w_E + \theta_K w_K$$

$$1 = w_C + w_G + (1 - \theta_E)w_E + (1 - \theta_K)w_K$$

4.1.3   Welfare

The marginal excess burden, *MEB*, can be derived along the lines of appendix 3B. It can be written as the sum of the following three market distortions:

$$MEB = \underset{\substack{labor-market \\ distortion}}{-\theta_L w_L \tilde{L}} \quad - \underset{\substack{capital-market \\ distortion}}{\theta_K w_K \tilde{K}} \quad - \underset{\substack{environmental \\ distortion}}{[\theta_E - \frac{u_M \gamma_E}{\lambda}]w_E \tilde{E}} \tag{4.5}$$

The second term on the RHS of (4.5) stands for the distortion in the capital market. In particular, the tax rate on capital drives a wedge between the marginal social benefits and costs of capital demand so that an expansion of capital demand raises welfare.

Table 4.2: Price elasticities of factor demands (conditional on employment) in the model with mobile capital

$Y=F[L,K,E]$	$Y = F[q(L,K),E]$	$Y = F[q(E,K),L]$	$Y = F[q(L,E),K]$
ϵ^*_{EL}	$-\dfrac{w_K \sigma_{LK} + w_L \sigma_E}{1 - w_E}$	$-\sigma_L$	$-\sigma_{EL}$
ϵ^*_{EE}	$(1 - w_E)\sigma_E$	$\dfrac{w_K \sigma_{EK} + w_E \sigma_L}{1 - w_L}$	σ_{EL}
ϵ^*_{EK}	$\dfrac{w_K(\sigma_{LK} - \sigma_E)}{1 - w_E}$	$\dfrac{w_K(\sigma_L - \sigma_{EK})}{1 - w_L}$	0
ϵ^*_{KL}	$-\sigma_{LK}$	$-\sigma_L$	$-\dfrac{w_E \sigma_{EL} + w_L \sigma_K}{1 - w_K}$
ϵ^*_{KE}	0	$\dfrac{w_E(\sigma_L - \sigma_{EK})}{1 - w_L}$	$\dfrac{w_E(\sigma_{EL} - \sigma_K)}{1 - w_K}$
ϵ^*_{KK}	σ_{LK}	$\dfrac{w_E \sigma_{EK} + w_K \sigma_L}{1 - w_L}$	σ_K

Alternatively, the marginal excess burden can be written as in (3.11), where the tax base effect is equal to:

$$\tilde{B} = \theta_L w_L \tilde{L} + \theta_E w_E \tilde{E} + \theta_K w_K \tilde{K} \tag{4.6}$$

Hence, the tax-base effect is determined not only by changes in the bases of the labor tax and the pollution tax, but also by changes in the capital tax base.

Substituting the government budget constraint (T4.7) into (4.6), we can rewrite the tax-base effect as in (3.13).

We can rewrite the effect on the labor-tax base in (4.6) in terms of the other two tax bases by first substituting the expression for labor supply (T4.6) into (4.6) and then using the government budget constraint (T4.7) and the non-profit condition (T4.4) to eliminate the wage rate. Thus, we find:

$$MEB = - \frac{1 - \theta_L}{1 - \theta_L - \theta_L \eta_{LL}} [\theta_E w_E \tilde{E} + \theta_K w_K \tilde{K}] + \frac{u_M \gamma_E}{\lambda} w_E \tilde{E}$$

(4.7)

$$\underset{dividend}{blue} \qquad\qquad \underset{dividend}{green}$$

Expression (4.7) shows that the blue dividend depends on the bases of the capital tax and the pollution tax alone. Indeed, the implications of the policy for the base of the labor tax are implicitly incorporated in (4.7). This chapter will frequently use (4.7) to analyze whether an environmental tax reform yields a strong double dividend.

4.2   Optimal taxation

This section derives the optimal tax system in the model with mobile capital. First, we find the optimum if all three tax rates can be adjusted. Next, the optimal tax structures under a fixed capital tax and a fixed labor tax are derived.

4.2.1   Unconstrained optimization

Appendix 4A solves the optimal tax problem if one can optimize with respect to all three taxes. This yields:[30]

$$\theta_K = 0$$

(4.8)

---

[30] See e.g. Auerbach (1985) for an introduction in the theory of optimal taxation.

$$\theta_E = \frac{1}{\eta} \frac{u_M \gamma_E}{\lambda} \qquad (4.9)$$

$$T_L = (1 - \frac{1}{\eta}) \frac{1}{\eta_{LL}} \qquad (4.10)$$

where $\eta = \lambda/\mu$ is the MCPF. The optimal tax rates on the polluting input and labor are equivalent to the optimal taxes in the benchmark model (compare (4.9) with (3.18) and (4.10) with (3.19)). The optimal tax on capital is zero. Without environmental externalities (i.e. $u_M = 0$), the optimal tax on the polluting input is zero as well. Intuitively, both capital and the polluting input are perfectly mobile internationally and thus supplied with an infinite elasticity. Accordingly, the incidence of both taxes is borne by the only immobile factor of production, labor. Hence, the capital tax and the pollution tax are implicit taxes on labor income. Just as in case of the benchmark model, a direct, explicit tax on labor is a more efficient instrument to raise revenues than an indirect, implicit tax on labor because the latter distorts not only the labor market but also other markets, such as the capital market and the market for polluting inputs. Consequently, in the absence of environmental considerations, the government finds it optimal to set zero taxes on both capital and the polluting input (see (4.8) and (4.9) if $u_M = 0$). If environmental considerations are taken into account, a pollution tax should internalize part of the environmental externality. In accordance with the targeting principle, this tax is levied on the polluting input only (see (4.9) if $u_M > 0$). Similar to the conclusions from the benchmark model, the optimal environmental tax is below the Pigovian level if the MCPF exceeds unity.

4.2.2   Optimal tax with fixed capital or pollution tax

We now derive the optimal taxes if the government can use only two tax rates to optimize welfare. In particular, we solve the optimal tax problem in case either T_L and T_E or T_L and T_K can be controlled, while the other tax is fixed at a possibly sub-optimal rate.

Appendix 4A derives the following solution for the optimal tax structure if, respectively, the capital tax and the pollution tax are fixed:

$$\theta_E = \frac{1}{\eta} \frac{u_M \gamma_E}{\lambda} - \frac{\Gamma_{EK}}{\Gamma_{EE}} \theta_K \qquad (4.11)$$

$$\theta_K = [\frac{1}{\eta} \frac{u_M \gamma_E}{\lambda} - \theta_E] \frac{\Gamma_{KE}}{\Gamma_{KK}} \qquad (4.12)$$

where $\Gamma_{ij} = \epsilon_{ij}^* - (w_j / w_L)\epsilon_{iL}^*$ for $i,j = E,K$. These Γ's denote the 'general equilibrium elasticities' that measure the consequences for the input mix due to a higher price of factor j and a lower price of labor induced by the rise in the price of input j along the factor-price frontier.[31] To illustrate, Γ_{EE} measures how the pollution/labor ratio is affected by a higher price of polluting inputs (measured by ϵ_{EE}^*) and a lower price of labor (measured by ϵ_{EL}^*). Similarly, Γ_{KE} measures the consequences for the capital/labor ratio due to a higher price of the polluting input (measured by ϵ_{KE}^*) and the associated lower price of labor (measured by ϵ_{KL}^*).

To obtain a better understanding of the factors determining these general equilibrium elasticities, Table 4.3 presents them in terms of Allen elasticities of substitution for three separable production functions. The table reveals that the own-price elasticities, Γ_{EE} and Γ_{KK}, are positive. In particular, both the higher price of pollution (capital) and the associated lower price of labor induce input substitution away from pollution (capital) towards labor. Furthermore, pollution (capital) involves the mobile factor (which is infinitely elastically in supply) whereas labor is the immobile factor (which is rather inelastic in supply). Accordingly, output unambiguously falls. This adverse production effect reinforces the fall in pollution (capital) on account of the substitution effect.

Table 4.3 reveals that Γ_{EK} and Γ_{KE} are also typically positive. Intuitively, the price of the immobile factor falls (which is rather inelastic in supply), while

---

[31] The production function exhibits constant returns to scale with respect to L, K and E. Hence, the factor-price frontier expresses producer wages in terms of the other input prices (see (T4.4)).

the price that is increased directly affects the mobile factor (which is infinitely elastic in supply). Consequently, output falls -- especially if the more expensive mobile factor is a good substitute for the immobile factor. The adverse production effect typically reduces the demand for pollution or capital.[32]

If we ignore environmental considerations (i.e. $u_M = 0$), the terms on the RHS of (4.11) and (4.12) are typically negative. Consequently, the polluting input should be subsidized if the capital tax is set exogenously at a positive value (see (4.11)). Intuitively, the pollution subsidy raises production, thereby expanding capital demand. In this way, the pollution subsidy acts as an indirect instrument to alleviate the distortion in the capital-market. A similar argument holds for the optimal capital tax if the tax rate on pollution is fixed (see (4.12)).

If environmental considerations are taken into account, the optimal pollution tax in (4.11) contains two components. First, environmental taxes aim at internalizing pollution externalities (see the first term on the RHS of (4.11)). This environmental term is smaller than the Pigovian tax if distortionary taxes raise the MCPF above unity. Intuitively, if public revenues are scarce, as reflected in a large value for the MCPF, the government cannot afford to fully internalize environmental externalities. Indeed, it relies less on taxes that are targeted at environmental protection (i.e. the pollution tax) and more on taxes that are efficient from a revenue-raising point of view (i.e. the labor tax). Whereas the first term thus offsets the environmental distortion, the second term on the RHS of (4.11) indicates that pollution taxes should be adopted also to alleviate the capital-market distortion that is due to the tax on capital. This latter task of the pollution tax calls for a subsidy rather than a tax.

---

[32] Γ_{KE} is negative only if positive substitution effects, associated with the change in relative prices, dominate this adverse production effect. Substitution effects stimulate the demand for capital if, compared to the cheaper factor (i.e. labor), the more expensive factor (the polluting input) is a better substitute for capital. The adverse production effect is small if the mobile factor that is increased in price (the polluting input) is a poor substitute for the immobile factor and if the immobile factor accounts for a large production share compared to the more expensive mobile factor (i.e. $\sigma_L < w_L \sigma_{EK}$). In the rest of this subsection, we assume that this is not the case so that $\Gamma_{KE} > 0$. For a similar reason, $\Gamma_{EK} > 0$.

Table 4.3: General equilibrium price elasticities of factor demands (conditional on employment) of a higher price of the polluting input (or capital) and a lower price of labor

	$Y=F[q(L,K),E]$	$Y=F[q(E,K),L]$	$Y=F[q(L,E),K]$
Γ_{EE}	$\dfrac{w_L\sigma_E + w_E w_K\sigma_{LK}}{w_L(1-w_E)}$	$\dfrac{w_E\sigma_L + w_L w_K\sigma_{EK}}{w_L(1-w_L)}$	$\dfrac{(1-w_K)\sigma_{EL}}{w_L}$
Γ_{KE}	$\dfrac{w_E}{w_L}\sigma_{LK}$	$\dfrac{w_E}{w_L}\dfrac{\sigma_L - w_L\sigma_{EK}}{1-w_L}$	$\dfrac{w_E}{w_L}\sigma_{EL}$
Γ_{EK}	$\dfrac{w_K}{w_L}\sigma_{LK}$	$\dfrac{w_K}{w_L}\dfrac{\sigma_L - w_L\sigma_{EK}}{1-w_L}$	$\dfrac{w_K}{w_L}\sigma_{EL}$
Γ_{KK}	$\dfrac{(1-w_E)\sigma_{LK}}{w_L}$	$\dfrac{w_K\sigma_L + w_L w_E\sigma_{EK}}{w_L(1-w_L)}$	$\dfrac{w_L\sigma_K + w_K w_E\sigma_{EL}}{w_L(1-w_K)}$
$\Gamma_{EE}-\Gamma_{KE}$	$\dfrac{\sigma_E - w_E\sigma_{LK}}{1-w_E}$	σ_{EK}	σ_{EL}
$\Gamma_{KK}-\Gamma_{EK}$	$\dfrac{w_E}{w_K}\sigma_{LK}$	$\dfrac{w_E}{w_K}\sigma_{EK}$	$\dfrac{w_E}{w_K}\dfrac{\sigma_K - w_K\sigma_{EL}}{1-w_K}$

Optimal tax structure if T_L is fixed

$\dfrac{\Gamma_{KK}-\Gamma_{EK}}{\Gamma_{EE}-\Gamma_{KE}}$	$1 - \dfrac{\sigma_E - \sigma_{LK}}{\sigma_E - w_E\sigma_{LK}}$	1	$1 - \dfrac{\sigma_{EL} - \sigma_K}{(1-w_K)\sigma_{EL}}$

With an exogenous pollution tax, the sign of the optimal capital tax corresponds to the sign of the term between square brackets on the RHS of (4.12). This term measures the net social damage due to an additional unit of pollution. Without a tax on polluting inputs, the social value of pollution is negative due to the associated adverse pollution externalities. In that case, by reducing production, a positive capital tax acts as an indirect instrument to internalize these externalities. If the pollution tax is sufficiently large, however, the net social value of an additional unit of pollution is positive, as the social value of the increase in tax revenue due to the broadening of the pollution tax base offsets the environmental costs. With additional pollution raising welfare, the optimal capital tax is negative; by boosting production, the capital subsidy expands the demand for the polluting input, thereby generating a first-order welfare gain.

4.2.3   Optimal tax with fixed labor tax

If the labor tax T_L is fixed at a level that yields insufficient revenues to meet the revenue requirement, positive tax rates on either capital or the polluting input are required. Let us now investigate how the optimal taxes T_K and T_E look like if the labor tax cannot be changed from its predetermined level.

Solving the optimal tax problem with respect to the taxes on capital and the polluting input, we find (see appendix 4A):[33]

$$\frac{\theta_E - \dfrac{1}{\eta}\dfrac{u_M \gamma_E}{\lambda}}{\theta_K} = \frac{\Gamma_{KK} - \Gamma_{EK}}{\Gamma_{EE} - \Gamma_{KE}} \tag{4.13}$$

The elasticities on the RHS of (4.13) can be expressed as follows:

$$\Gamma_{EE} - \Gamma_{KE} = \epsilon_{EE}^* - \frac{w_E}{w_K}\epsilon_{EK}^* \tag{4.14}$$

---

[33] Expression (4.13) is similar to the optimal tax formula between labor taxes and pollution taxes in Bovenberg and van der Ploeg (1996) if in our model $\eta_{LL} = 0$. Their model contains two inputs that are supplied with infinite elasticity (i.e. polluting resources and labor) and one fixed factor. In our model, labor plays the role of the fixed factor if it is supplied inelastically.

$$\Gamma_{KK} - \Gamma_{EK} = \epsilon_{KK}^* - \frac{w_K}{w_E} \epsilon_{KE}^* \tag{4.15}$$

The elasticity in (4.14) measures how the pollution/labor ratio changes due to a higher producer price of the polluting input (measured by ϵ_{EE}^*) and the associated lower price of capital (measured by ϵ_{EK}^*). Table 4.3 presents this general equilibrium elasticity for three separable production functions. It reveals that a higher price of the polluting input and a lower price of capital typically reduce the pollution/labor ratio, i.e. $\Gamma_{EE} - \Gamma_{EK} > 0$. We refer to this as the normal case. In the exceptional case that the term in (4.14) is negative, a higher price of polluting inputs, rather paradoxically, raises pollution. Intuitively, the higher price of polluting inputs allows for a lower price of capital. This raises production and thereby pollution. In that case, compared to a pollution tax, a capital tax is a better instrument to cut pollution.[34] This exceptional case occurs if, compared to pollution, capital is a much better substitute for labor (see Table 4.3 if E is separable and σ_E is small compared to σ_{LK}). The tax reform then boosts production because the mobile factor that is elastic in demand (i.e. capital which is a good substitute for labor) becomes cheaper, while the mobile factor that is rather inelastic in demand (i.e. the polluting input, which is a relatively poor substitute for labor) becomes more expensive. At the same time, the adverse substitution effect on the demand for the polluting input due to the higher cost of pollution is small because the polluting input is a poor substitute for the two other inputs (i.e. σ_E is relatively small). Accordingly, the positive output effect dominates the adverse substitution effect and pollution thus increases.

The general equilibrium elasticity in (4.15) measures how capital demand (scaled by the immobile input, labor) changes as a result of a higher price for the polluting input (measured by ϵ_{KK}^*), and the lower price of capital (measured by

---

[34] Schöb (1996) finds a similar result for pollution taxes on consumption. In particular, he shows that a tax on a non-polluting consumption commodity that is complementary to the polluting consumption commodity may be a better instrument for environmental protection than a tax on the polluting commodity.

ϵ_{KE}^{*}). Table 4.3 shows that the elasticity in (4.15) is typically positive, so that the demand for capital expands. Intuitively, lower capital costs boost the demand for capital through positive substitution effects. In the normal case, this positive substitution effect dominates the arbitrary output effect. Hence, capital demand increases and private welfare rises. Only in the exceptional case that a negative output effect is strong enough to offset the positive substitution effect does the demand for capital decline. This occurs if the polluting input is a much better substitute for labor than capital (see Table 4.3 if K is separable and σ_K is small compared to σ_{EL}). Intuitively, output contracts because the tax burden is shifted away from the input that is inelastic in demand (i.e. capital, which is a poor substitute for labor) towards the input that is in relatively elastic demand (i.e. the polluting input, which is a relatively good substitute for labor). With only limited possibilities to substitute capital for other inputs, the adverse production effect dominates the substitution effect so that capital demand drops.

The LHS of expression (4.13) presents the ratio of the environmental tax -- corrected for the social value of the external effect of pollution -- and the capital tax. The RHS of (4.13) reveals that this optimal-tax ratio depends on the general equilibrium demand elasticities. The last row of Table 4.3 presents this optimal tax ratio for three separable production functions. It reveals that the polluting input and capital should both be taxed if these two inputs do not differ too much in the ease with which they can be substituted for labor. To illustrate, capital and the polluting input are equally good substitutes for labor if they enter the same nest of the CES production function. In that case, it is optimal to levy uniform taxes on capital and the polluting input. If capital and pollution do not enter the same nest in the CES production structure, the highest tax should be levied on the factor that is the poorest substitute for labor. Intuitively, poor substitution with the immobile factor render the demand for the factor inelastic. In accordance with the Ramsey principle of optimal taxation, inelastic demands should be taxed relatively heavily to limit tax distortions. If substitution possibilities with labor differ substantially across the two mobile factors, the factor that is the best substitute for labor (i.e. the factor that is elastic in demand) should be subsidized. To illustrate, if, compared to capital, pollution is a much better substitute for labor, it may be optimal to subsidize, rather than tax, pollution (see Table 4.3 if K is separable and σ_{EL} is much larger than σ_K).

By using the first-order conditions in appendix 4A, we can eliminate the capital tax from (4.13) and arrive at the following reduced-form for the optimal pollution tax:

$$\theta_E = \frac{1}{\eta}\frac{u_M \gamma_E}{\lambda} + \frac{1}{1+T_L}[(1-\frac{1}{\eta})\frac{1}{\eta_{LL}} - T_L]\frac{\dfrac{w_L(\Gamma_{KK}-\Gamma_{EK})}{\Gamma_D}}{\dfrac{1}{\eta_{LL}} + \dfrac{[w_K(\Gamma_{EE}-\Gamma_{KE})+w_E(\Gamma_{KK}-\Gamma_{EK})]}{\Gamma_D}} \qquad (4.16)$$

where $\Gamma_D = \Gamma_{KK}\Gamma_{EE} - \Gamma_{EK}\Gamma_{KE}$. The optimal pollution tax consists of two terms: the first term on the RHS of (4.16) represents the role of pollution taxes in internalizing environmental externalities. Just as in (4.11), the weight that the government assigns to internalizing externalities declines with the MCPF. Second, if the government cannot freely use the labor tax -- which is the most efficient instrument to raise revenues -- the government may wish to employ pollution taxes as a revenue-raising device. The second term on the RHS of (4.16) reflects this revenue-raising task of pollution taxes. This term is zero if the labor tax is at its optimal level, as is indicated by the term between square brackets on the RHS of (4.16). Hence, the optimal pollution tax would correspond to its level in the optimal tax system (see (4.9) and (4.10)). However, if the initial labor tax is below its optimal level, the term between square brackets is positive. In that case, labor taxes do not raise sufficient revenues; pollution and capital taxes are thus necessary to meet revenue requirements. The optimal pollution tax may thus exceed the Pigovian level. In accordance with Ramsey principles, pollution should be taxed relatively heavily if, compared to capital, pollution is inelastic in demand.

4.3   Tax reform from labor to pollution taxes

This section explores the consequences of a rise in pollution taxes if the revenues are used to cut labor taxes. We discuss the effects of the environmental tax reform starting from different initial equilibria. Appendix 4B derives the solutions of a rise in pollution taxes, recycled through lower labor taxes. The reduced-form coefficients are presented in the second column of Table 4.4.

Table 4.4: Reduced-form coefficients for the pollution tax in the model with mobile capital and an endogenous labor tax

	$\tilde{T}_E$
$\Delta_L \tilde{W}$	$-[\theta_E \Gamma_{EE} + \theta_K \Gamma_{EK}] w_E$
$\Delta_L \tilde{I}.$	$-[\theta_E \Gamma_{EE} + \theta_K \Gamma_{EK}] w_E \eta_{LL}$
$\Delta_L \tilde{E}$	$-[\Gamma_{EE} \Delta_L + (\theta_E \Gamma_{EE} + \theta_K \Gamma_{EK}) w_E \eta_{LL}]$
$\Delta_L \tilde{K}$	$-[\Gamma_{KE} \Delta_L + (\theta_E \Gamma_{EE} + \theta_K \Gamma_{EK}) w_E \eta_{LL}]$
$\Delta_L \tilde{Y}$	$-[(\Delta_L + \theta_E \eta_{LL}) \Gamma_{EE} + (\Delta_L + \theta_K \eta_{LL}) \Gamma_{EK}] w_E$

$$\Delta_L = (1 - \theta_L) w_L - \eta_{LL}[\theta_L w_L + \theta_E w_E + \theta_K w_K]$$

4.3.1   Starting from zero non-labor taxes

If the initial tax rates on the polluting input and capital are zero (i.e. $\theta_E = \theta_K = 0$), the reduced-form coefficient in the first row of Table 4.4 reveals that introducing a pollution tax leaves wages and thus blue welfare unaffected. Hence, zero taxes on the polluting input and capital are optimal from a non-environmental point of view and thus optimize blue welfare (see also (4.8) - (4.10)). A marginal change in the tax system, therefore, does not yield any first-order effects on private welfare. Green welfare improves, however, due to higher pollution taxes as pollution falls (since $\Gamma_{EE} > 0$).

4.3.2   Starting from a positive pollution tax

If we start from zero taxes on capital (i.e. $\theta_K = 0$) and a positive pollution tax (i.e. $\theta_E > 0$), further increases in the pollution tax amount to a shift in the tax mix away from the non-environmental optimum (in which $\theta_E = \theta_K = 0$). The resulting decline in non-environmental welfare implies a fall in private incomes and thus a negative blue dividend. The reduced-form coefficient in the first row of Table 4.4 indeed reveals that wages (representing blue welfare) fall due to a green tax shift if the pollution tax is positive initially. The reason is that, at the margin, pollution taxes are less efficient instruments to raise revenue than taxes on labor because they not only distort the labor market but also reduce the demand for the polluting input, thereby eroding the base of the pollution tax. The associated drop in private incomes measures the costs of a cleaner environment, i.e. of the green dividend. These private abatement costs reduce wages, thereby reducing the incentives to supply labor. The environmental benefits, in contrast, leave labor supply unaffected: these benefits are public and thus independent of labor supply. Indeed, by improving the quality of the environment, pollution taxes expand the provision of public consumption goods. As in chapter 3, we define the *tax-burden effect* as the higher tax burden associated with the additional costs due to a higher overall provision of public goods. The erosion of the base of the pollution tax measures this tax-burden effect.

The magnitude of the tax-burden effect is determined by two elements: the initial tax rate on the polluting input and the fall in the demand for polluting inputs. The initial pollution tax measures the marginal costs of improving environmental quality. In particular, if pre-existing pollution taxes are large, a decline in the demand for the polluting input substantially reduces public revenues by eroding the tax base.

The second element determining the tax-burden effect is the general equilibrium elasticity of polluting input demand, Γ_{EE}, defined in Table 4.3. A large elasticity, on the one hand, implies that the environmental tax reform is successful in cutting pollution. On the other hand, in that case a higher pollution tax imposes a heavy burden on the private sector, as tax revenues fall substantially due to an eroding tax base. Hence, a trade-off exists between green and blue welfare: if an environmental tax reform is successful in enhancing environmental quality, it imposes a high burden on the private sector in terms of lower wages.

4.3.3   Starting from a positive capital tax

If the initial equilibrium features a zero pollution tax (i.e. $\theta_E = 0$) and a positive tax rate on clean capital (i.e. $\theta_K > 0$), the tax system is sub-optimal, even from a non-environmental point of view. In that case, substituting pollution taxes for labor taxes may either exacerbate or alleviate pre-existing tax distortions. In particular, an environmental tax reform reduces wages if capital demand falls (see the first row in Table 4.4 with $\theta_E = 0$). The effect on capital demand is determined by the cross-price general equilibrium elasticity of capital demand, i.e. $\Gamma_{EK} = (w_K / w_E)\Gamma_{KE}$ (see the reduced-form for capital in the fourth row of Table 4.4 with $\theta_E = 0$). This elasticity measures the consequences for the capital/labor ratio due to a higher price of the polluting input and the associated lower price of labor. Table 4.3 reveals that capital demand typically falls. This hurts welfare by eroding the base of the tax on capital. In this way, substituting pollution taxes for labor taxes exacerbates pre-existing distortions associated with capital taxation. We call this the *tax-shifting effect* because the tax burden is shifted towards capital in terms of a lower capital demand. The tax-shifting effect is measured by the second term between square brackets on the RHS of (4.7).[35]

The magnitude of the tax-shifting effect is determined by the initial tax rate on capital and the cross-price elasticity of capital demand. High initial taxes on capital imply that the social costs associated with less capital demand are large. In particular, a fall in the demand for capital induces a large erosion of the tax base and, therefore, a substantial fall in private incomes. Second, the cross-price elasticity of capital demand, Γ_{KE}, determines how large the fall in capital demand will be.

4.3.4   Starting from an arbitrary equilibrium

We have now uncovered three channels through which a tax shift away from labor towards pollution affects welfare. First, substitution away from polluting inputs in combination with an adverse production effect contributes to green welfare (i.e. the final term on the RHS of (4.7)). Second, pre-existing taxes on the polluting input cause a tax-burden effect. In particular, the fall in pollution raises the tax burden on private incomes, thereby adversely affecting blue welfare (see the first term between square brackets on the RHS of (4.7)). Third, if the capital income tax is initially positive, an environmental tax reform induces a tax-shifting effect by affecting the base of the capital income tax. Through this channel, the tax distortion on the capital market is typically exacerbated, so that blue welfare falls (see the second term between square brackets on the RHS of (4.7)). Overall, starting from an arbitrary equilibrium in which both capital and pollution taxes are positive, an environmental tax reform

---

[35] The ultimate burden of all taxes in terms of reduced real incomes is borne by labor because the supply of capital is infinitely elastic. Thus, the burden of taxation on the capital market is reflected only in a decline of capital demand and not a drop in the after-tax return on capital. The burden in terms of reduced incomes is shifted back onto labor. In a more general model in which capital supply would be less than infinitely elastic, the burden of taxation in the capital market would reduce not only capital demand but also the after-tax return on capital. Hence, labor and capital would share the burden of taxation in terms of lower after-tax incomes (see also section 4.5 - 4.7). In that case, the tax-shifting effect involves a redistribution of the tax burden across incomes, rather than a decline in capital demand.

yields a green dividend. However, blue welfare typically falls. Hence, there is no strong double dividend.

4.4 Tax reform from capital to pollution taxes

This subsection explores the consequences of shifting the tax burden away from capital taxes, rather than from taxes on labor. The solutions are presented in the second column of Table 4.5 (see appendix 4B for a derivation).

4.4.1 Starting from zero non-labor taxes

If taxes on polluting inputs and capital are zero (i.e. $\theta_E = \theta_K = 0$) a rise in pollution taxes leaves the wage rate unaffected (see the reduced form in the first row of Table 4.5 with $\theta_E = \theta_K = 0$). Indeed, as we start from a tax system that is optimal from a non-environmental point of view (see section 4.2), a marginal change in the tax system away from this optimum does not yield any first-order effects on blue welfare.

4.4.2 Starting from a positive pollution tax

If the initial equilibrium features a positive pollution tax (i.e. $\theta_E > 0$) and zero taxes on capital (i.e. $\theta_K = 0$), further increases in the pollution tax typically shift the tax system further away from the non-environmental optimum, thereby harming blue welfare. Indeed, the reduced-form coefficient in the first row of Table 4.5 reveals that wages decline if the initial pollution tax is positive and the demand for polluting inputs falls. The effect on pollution is determined by $\Gamma_{EE} - \Gamma_{KE}$ (see the third row in Table 4.5 with $\theta_E = \theta_K = 0$), which, for three separable production functions, is defined in Table 4.3. We find that the shift from capital towards pollution taxes typically yields a green dividend by reducing pollution. In this so-called 'normal case,' the reduction in pollution erodes the base of the environmental tax and, given an exogenous revenue-requirement, the government needs to raise other tax rates, thereby harming private welfare. In the exceptional case that the term $\Gamma_{EE} - \Gamma_{KE} < 0$, the introduction of a pollution tax, rather paradoxically, raises pollution. In that case, the tax reform yields a positive blue dividend by expanding the base of the pollution tax -- but at the cost of a negative green dividend. Thus, the double dividend always fails if the initial equilibrium features a positive pollution tax and a zero capital tax: the blue and green dividends have opposite signs.

Table 4.5: Reduced-form coefficients for the pollution tax in the model with mobile capital and an endogenous capital tax

	$\tilde{T}_E$
$\Delta_K \tilde{W}$	$-[\theta_E(\Gamma_{EE} - \Gamma_{KE}) - \theta_K(\Gamma_{KK} - \Gamma_{EK})]w_E$
$\Delta_K \tilde{L}$	$-[\theta_E(\Gamma_{EE} - \Gamma_{KE}) - \theta_K(\Gamma_{KK} - \Gamma_{EK})]w_E\eta_{LL}$
$\Delta_K \tilde{E}$	$-(\Gamma_{EE} - \Gamma_{KE})\Delta_K$ $-[\theta_E(\Gamma_{EE} - \Gamma_{KE}) - \theta_K(\Gamma_{KK} - \Gamma_{EK})]w_E\eta_{LL}$
$\Delta_K \tilde{K}$	$\dfrac{w_E}{w_K}(\Gamma_{KK} - \Gamma_{EK})\Delta_K$ $-[\theta_E(\Gamma_{EE} - \Gamma_{KE}) - \theta_K(\Gamma_{KK} - \Gamma_{EK})]w_E\eta_{LL}$
$\Delta_K \tilde{Y}$	$-(\Delta_K + \theta_E\eta_{LL})(\Gamma_{EE} - \Gamma_{KE})w_E$ $+(\Delta_K + \theta_K\eta_{LL})(\Gamma_{KK} - \Gamma_{EK})w_E$

$\Delta_K = [(1 - \theta_L)w_L - \theta_L w_L \Gamma_{KE} - \theta_K w_L \Gamma_{KK}] - \eta_{LL}[\theta_L w_L + \theta_E w_E + \theta_K w_K]$

4.4.3   Starting from a positive capital tax

Substituting pollution taxes for capital taxes is likely to produce a double dividend if such a reform starts from an initial equilibrium with a zero pollution tax (i.e. $\theta_E = 0$) and a positive tax on capital (i.e. $\theta_K > 0$). With this initial equilibrium, expression (4.7) reveals that the sign of the blue dividend depends on how the reform affects capital demand. This effect is determined by the general equilibrium elasticity, $\Gamma_{KK} - \Gamma_{EK}$ (see the fourth row in Table 4.5 with $\theta_E = 0$). The intuition behind the expansion of private welfare is that the environmental tax reform moves the tax system closer to the non-environmental optimum. In particular, in the initial equilibrium, one of the non-polluting factors, namely capital, is overtaxed compared to the other non-polluting factor, namely, labor. Raising the pollution tax and reducing the capital tax typically reduces the tax burden on the overtaxed factor because, in contrast to the pollution tax, the capital tax directly impacts the capital market. The reallocation of the tax burden away from the overtaxed factor, resulting in an expansion of capital demand, is called the tax-shifting effect.

Only in the exceptional case that a negative output effect is strong enough to offset the positive substitution effect does the demand for capital decline so that private welfare falls. In this case, the tax-shifting effect is negative, since it works in the wrong way by further raising the burden of taxation on the capital market.

4.4.4   Starting from an arbitrary equilibrium

The welfare effects of an ecological tax reform starting from an arbitrary equilibrium can be decomposed into three parts. First, in the normal case, substitution effects produce a green dividend (see final term on the RHS of (4.7)). Second, with pre-existing taxes on the polluting input, the base of the environmental taxes erodes in the normal case, implying a positive tax-burden effect. This reduces blue welfare (see first term between the square brackets on the RHS of (4.7)). Third, with pre-existing taxes on capital income, a positive tax-shifting effect creates the potential for an improvement in private welfare (see the second term between the square brackets on the RHS of (4.7)).

In the normal case, substitution effects dominate output effects so that the demand for capital expands while the demand for the polluting input contracts. Consequently, both the tax-shifting effect and the tax-burden effect are positive. Blue welfare improves if the tax-shifting effect is large compared to the tax-burden effect, so that the improvement in non-environmental efficiency of the tax system (i.e. the tax-shifting effect) 'finances' the costs of enhancing environmental quality (i.e. the tax-burden effect). A double dividend requires that the initial capital tax is large compared to the initial pollution tax. Intuitively, large initial capital taxes indicate substantial potential for non-environmental efficiency improvements, while small initial pollution taxes imply that environmental improvements are relatively cheap.

A double dividend is feasible only in the normal case, i.e. if substitution effects dominate output effects. In the exceptional case that large positive output effects cause pollution to rise, the green dividend is negative. In the other exceptional case, the blue dividend is negative, as adverse production effects reduce capital demand.

4.5   The model with fixed capital in production

Whereas sections 4.1 - 4.4 discuss the extreme case in which capital is perfectly mobile internationally and thus supplied with an infinite elasticity, this section introduces capital that is immobile internationally. Hence, the model contains two factors of production that cannot cross international borders (i.e. labor and capital) and one input that is perfectly mobile across countries (i.e. the polluting input). Concerning capital, we take the extreme case in which it is fixed in supply -- contrary to labor which is supplied with a finite wage elasticity.

4.5.1   Structure of the model

The production function $F(L,E,H)$ contains three inputs, where H is a fixed factor owned by domestic households. The production function exhibits constant returns to scale with respect to these three inputs. Whereas the supply of the polluting input is infinitely elastic, the supply of the fixed factor is completely inelastic. The fixed factor can be interpreted as land or immobile capital.

In contrast to the model in section 4.1, the presence of a fixed factor allows for profits to exist. These profits (Π) can be interpreted as the return on the fixed factor. Hence, instead of a non-profit condition, we have the following profit equation:

$$\Pi = P_Y Y - (1 + T_L)WL - (P_E + T_E)E \qquad (4.17)$$

Domestic households own the firms so that they receive the profit income. The household budget available for consumption thus consists of an additional component compared to the benchmark model, namely, after-tax profits ($(1-T_\Pi)\Pi$):

$$WL + (1 - T_\Pi)\Pi = P_C C \qquad (4.18)$$

where T_Π denotes the profit tax. This additional tax implies that the government budget changes as follows:

$$P_G G = T_L WL + T_E E + T_\Pi \Pi \qquad (4.19)$$

4.5.2   Linearization

Table 4.6 presents the model in relative changes in the presence of fixed capital (as far as it differs from the benchmark model). Relations (T4.12) and (T4.13) represent the factor-demand equations for labor and the polluting input. The elasticities, ϵ_{ij}, $(i,j = L,K,E)$ represent price elasticities of factor demand, conditional on the level of the fixed factor. Similar to the previous section, we can -- for three separable production functions -- express these price elasticities in terms of shares and substitution elasticities. Comparing ϵ_{ij} with the Γ's in Table 4.3, we find that they look very similar. Indeed, the elasticities ϵ_{ij} have a similar interpretation as those in Table 4.3. However, whereas the Γ's in Table 4.3 are defined conditional on the level of employment, the ϵ's are defined conditional on the fixed factor. In view of the similarity between the elasticities, this section does not present the elasticities for different separable production functions.[36] Moreover, this section does not discuss in detail the role of factor substitution.

Expression (T4.16) reveals that a rise in the after-tax wage rate (W) boosts labor supply if the uncompensated wage elasticity (η_{LL}) is positive. In addition, profit income allows for an income effect. Indeed, depending on the income elasticity of labor supply (η_L), a rise in after-tax profits reduces labor supply because of a negative income effect.

Table 4.6: Linearized model with fixed capital

Firms

Domestic output	$\tilde{Y} = w_H\tilde{H} + w_E\tilde{E} + w_L\tilde{L}$	(T4.11)
Polluting input demand	$\tilde{E} = \tilde{H} - \epsilon_{EL}(\tilde{W} + \tilde{T}_L) - \epsilon_{EE}\tilde{T}_E$	(T4.12)
Labor demand	$\tilde{L} = \tilde{H} - \epsilon_{LL}(\tilde{W} + \tilde{T}_L) - \epsilon_{LE}\tilde{T}_E$	(T4.13)
Profits	$(1 + \theta_\Pi)w_\Pi\tilde{\Pi} = w_H\tilde{H} - w_L(\tilde{W} + \tilde{T}_L) - w_E\tilde{T}_E$	(T4.14)

Households

Household budget	$w_L(1 - \theta_L)(\tilde{L} + \tilde{W}) + w_\Pi(\tilde{\Pi} - \tilde{T}_\Pi) = w_C\tilde{C}$	(T4.15)
Labor supply	$\tilde{L}_S = \eta_{LL}\tilde{W} - \eta_L(w_\Pi/w_C)(\tilde{\Pi} - \tilde{T}_\Pi)$	(T4.16)

---

[36] See Table 7.2 in chapter 7 for an exposition of these elasticities.

Government budget
$$0 = w_L \tilde{T}_L + w_E \tilde{T}_E + w_\Pi \tilde{T}_\Pi +$$
$$\theta_L w_L (\tilde{W} + \tilde{L}) + \theta_E w_E \tilde{E} + \theta_\Pi w_\Pi \tilde{\Pi}$$
(T4.17)

Labor-market equilibrium $\tilde{L}_S = \tilde{L}$ (T4.18)

Environmental quality $w_M \tilde{M} = - \gamma_E w_E \tilde{E}$ (T4.19)

Material balance $\tilde{Y} = w_C \tilde{C} + (1 - \theta_E) w_E \tilde{E}$ (T4.20)

Notation

E = demand for polluting inputs $\quad H$ = fixed capital

Y = output $\qquad \Pi$ = profits

L = employment $\qquad W$ = market wage

T_L = ad-valorem tax on labor income T_Π = tax on profits

T_E = tax on polluting inputs $\qquad C$ = consumption of clean goods

L_S = labor supply $\qquad M$ = environmental quality

Parameters

ϵ_{IJ} = demand elasticity (conditional on the fixed factor) of factor I with respect to changes in the price of factor J, for $I,J = E,L$

η_L = income elasticity of labor supply

η_{LL}^* = $\eta_{LL} + ((1-\theta_L)w_L/w_C)\eta_L$ compensated labor supply elasticity

γ_E = $-m_E / (P_E + T_E)$

Taxes

$$\tilde{T}_L = \frac{dT_L}{1 + T_L} \quad ; \quad \tilde{T}_E = \frac{dT_E}{P_E + T_E} \quad ; \quad \tilde{T}_\Pi = \frac{dT_\Pi}{1 - T_\Pi}$$

$$\theta_L = \frac{T_L}{1 + T_L} \quad ; \quad \theta_E = \frac{T_E}{P_E + T_E} \quad ; \quad \theta_\Pi = \frac{T_\Pi}{1 - T_\Pi}$$

Shares

$$w_L = (1+T_L)WL/P_YY \qquad w_E = (P_E+T_E)E/P_YY \qquad w_\Pi = (1-T_\Pi)\Pi/Y$$
$$w_H = (1+\theta_\Pi)w_\Pi \qquad\quad w_C = P_CC/P_YY \qquad\qquad w_G = P_GG/P_YY$$
$$w_M = M/P_YY$$

Relationships between shares

$$w_E + w_L + w_H = 1 \qquad\qquad\qquad (1 - \theta_L)w_L + w_\Pi = w_C$$

$$w_G = \theta_L w_L + \theta_E w_E + \theta_\Pi w_\Pi \qquad\qquad 1 = w_C + w_G + (1 - \theta_E)w_E$$

4.5.3   Welfare

The marginal excess burden, *MEB*, can be derived as in appendix 3B:

$$MEB = -[(1 - \theta_L)w_L\tilde{W} + w_\Pi(\tilde{\Pi} - \tilde{T}_\Pi)] - \frac{u_M}{\lambda}w_M\tilde{M} \tag{4.20}$$

$$\underbrace{\qquad\qquad\qquad\qquad\qquad}_{blue\ dividend} \qquad \underbrace{\qquad\qquad}_{green\ dividend}$$

Hence, blue welfare is no longer determined by wages alone. In particular, the *MEB* in (4.20) reveals that the effect on blue welfare depends on both after-tax wage income and after-tax profits.

4.6   Optimal taxation

This section explores the optimal tax structure in the model with fixed capital. First, it derives the optimal tax system if the government can adjust all taxes (i.e. the labor tax, the pollution tax and the profit tax). Then, it derives the optimal tax structure if either of the three taxes is fixed at a possibly sub-optimal level and the government optimizes with respect to the other two tax rates.

4.6.1   Unconstrained optimization

Appendix 4C derives the optimal tax structure if the government optimizes welfare with respect to the three tax rates, i.e. the labor tax, the pollution tax and the profit tax. If no restrictions are imposed on tax rates -- which may thus exceed 100% -- this yields the following solution:

$$T_L = 0 \tag{4.21}$$

$$\theta_E = \frac{u_M \gamma_E}{\lambda} \tag{4.22}$$

Expressions (4.21) and (4.22) reveal that the government finds it optimal from a non-environmental point of view (i.e. if $u_M = 0$) to set both the labor tax rate and the pollution tax rate at zero. Hence, all revenues are raised by profit taxes. Intuitively, the profit tax amounts to a non-distortionary or lump-sum tax. Hence, we are in a first-best world. Without environmental considerations, it is indeed optimal in such a first-best framework to raise all revenues through lump-sum profit taxes and set distortionary tax rates on labor and pollution at zero. In the presence of environmental externalities, the tax system aims not only at raising revenues with the least costs to private incomes, but also at internalizing the external effect of pollution. In accordance with the targeting principle, this calls for a positive tax on pollution (see (4.22) with $u_M > 0$). Note that the MCPF is equal to one in this first-best world (i.e. $\eta = 1$) so that the optimal pollution tax is equal to the Pigovian rate.

4.6.2   Optimal tax with a fixed profit tax[37]

Suppose that the profit tax is fixed at some predetermined rate and this tax raises insufficient revenues to meet the government revenue requirement. The government, therefore, needs to adopt distortionary taxes on labor and pollution to raise sufficient revenues. In that case, the optimal structure between labor and pollution taxes looks as follows (see appendix 4C for a derivation):

$$\frac{\theta_E - \frac{1}{\eta}\frac{u_M \gamma_E}{\lambda}}{\theta_L} = [\, 1 + (1 - \frac{1}{\eta})\frac{1}{\eta_{LL}} \,]\, \frac{\frac{(1 - T_\Pi)(\epsilon_{LL} - \epsilon_{EL})}{\epsilon_D}}{\frac{(1 - T_\Pi)(\epsilon_{EE} - \epsilon_{LE})}{\epsilon_D} + \frac{1}{\eta_{LL}^*}} \qquad (4.23)$$

where η_{LL}^* denotes the compensated wage elasticity of labor supply. The LHS of relation (4.23) presents the optimal ratio between pollution taxes -- corrected for the environmental externality -- and labor taxes. The RHS of (4.23) indicates that the optimal-tax ratio depends on demand and supply elasticities of labor and the polluting inputs. In particular, pollution should be taxed heavily compared to labor if labor demand and supply are elastic, relative to pollution demand and supply. The demand elasticities on the RHS of (4.23) are presented for three separable production functions in Table 4.7.

To gain more intuition behind the optimal tax structure in (4.23), we explore two special cases, namely the case in which the profit tax rate is 100% and the case in which labor is supplied infinitely elastically.

---

[37] Ligthart and van der Ploeg (1997) investigate a similar problem. However, they assume that labor supply does not respond to changes in profits. Furthermore, they do not derive an analytical solution for the optimal tax structure between polluting inputs and labor.

Table 4.7: Price elasticities of factor demands in the model with fixed capital under different separability assumptions

	$Y = F[q(L,H),E]$	$Y = F[q(E,H),L]$	$Y = F[q(L,E),H]$
$\epsilon_{LL} - \epsilon_{EL}$	σ_{LH}	$\dfrac{\sigma_L - w_L\sigma_{EH}}{1 - w_L}$	σ_{LE}
$\epsilon_{EE} - \epsilon_{LE}$	$\dfrac{\sigma_E - w_E\sigma_{LH}}{1 - w_E}$	σ_{EH}	σ_{LE}
ϵ_D	$\dfrac{\sigma_{LH}\sigma_E}{w_H}$	$\dfrac{\sigma_L\sigma_{EH}}{w_H}$	$\dfrac{\sigma_H\sigma_{LE}}{w_H}$

Profits cannot bear the burden of taxation

Suppose that the government optimizes the tax system with respect to the three tax rates, but that the profit tax is not allowed to exceed a 100% rate. In that case, the government will find it optimal to fully tax away profit incomes through a 100% profit tax (i.e. $T_\Pi = 1$). If this 100% tax on profits yields insufficient revenues to meet the government budget constraint, the optimal taxes on the other two inputs would look as follows:

$$\theta_E = \frac{1}{\eta}\frac{u_M \gamma_E}{\lambda} \qquad (4.24)$$

$$T_L = (1 - \frac{1}{\eta})\frac{1}{\eta_{LL}} \qquad (4.25)$$

Relations (4.24) and (4.25) are similar to the optimal tax formulas in the benchmark model in which the fixed factor is absent (i.e. $w_H = 0$).[38] Intuitively, if $T_\Pi = 1$, the government, rather than households, receives the return on fixed capital. Hence, households rely on labor incomes alone. Therefore, just as in the

---

[38] Expressions (4.24) and (4.25) will hold also if the share of the fixed factor, w_H, is infinitely small so that $\epsilon_D \to \infty$.

benchmark model, the incidence of both labor taxes and pollution taxes will be borne by these labor incomes. As direct labor taxes are more efficient instruments to raise revenue than pollution taxes, the latter taxes should not be adopted as revenue-raising instruments but only to internalize the environmental externality.

Labor supplied infinitely elastically

If labor would be supplied infinitely elastically (i.e. $\eta_{LL} \rightarrow \infty$), the model of this section would be similar to the model of section 4.1 (namely, if in that section labor would be supplied inelastically). In particular, the model of this section would contain two inputs that are supplied with an infinite elasticity (labor and the polluting input) and one input that is in fixed supply (fixed capital).[39] For a given tax rate on the fixed factor, the optimal tax rates on the two elastic factors would then be characterized by:

$$\frac{\theta_E - \dfrac{1}{\eta}\dfrac{u_M \gamma_E}{\lambda}}{\theta_L} = \frac{\epsilon_{LL} - \epsilon_{EL}}{\epsilon_{EE} - \epsilon_{LE}} \tag{4.26}$$

The optimal tax structure in (4.26) looks similar to (4.13). In particular, the two elastic inputs should be taxed according to their demand elasticities. Indeed, in accordance with the Ramsey principle for taxation, the government should impose the largest tax on the input that is relatively inelastic in demand (see also the discussion about the elasticities in section 4.2).[40]

General case

If labor is supplied with a finite elasticity, the denominator of the RHS of (4.23) contains also the reciprocal of the elasticity of labor supply. Hence, apart from

---

[39] In that case, labor would play the same role as capital does in section 4.1. The fixed factor is then comparable to fixed labor supply in section 4.1.

[40] In the exceptional case that pollution increases due to an environmental tax reform (see Table 4.7 if E is separable and $\sigma_E < w_E \sigma_{LH}$), it would be optimal to subsidize rather than to tax labor.

demand elasticities, also the labor supply elasticity matters for the optimal tax structure. In the numerator of (4.23), the reciprocal of the elasticity of pollution supply does not appear because the reciprocal of the infinite supply elasticity is zero.

4.6.3   Optimal tax system with fixed pollution tax

Suppose that the pollution tax is fixed at a possibly suboptimal level. Hence, the government cannot optimize with respect to the most direct instrument to internalize the environmental externality. It does, however, have access to lump-sum profit taxes which render the MCPF equal to unity. Furthermore, the government can use labor taxes to optimize the tax system. In particular, from the first-order conditions in appendix 4C, we derive for the optimal tax on labor:

$$
\theta_L = - \frac{(\theta_E - \frac{u_M \gamma_E}{\lambda}) \epsilon_{LE}}{\epsilon_{LL}}
\tag{4.27}
$$

The optimal labor tax in (4.27) is positive if the environmental tax is too small compared to the value of the external costs of pollution (i.e. $\theta_E < u_M \gamma_E / \lambda$). In that case, the government will find it optimal to adopt the labor tax as an instrument to indirectly internalize the environmental externality. Indeed, labor taxes typically reduce output and thus pollution. In this way, the labor tax corrects for the too high level of pollution. If the pollution tax exceeds the external damage costs (i.e. $\theta_E > u_M \gamma_E / \lambda$), an additional unit of pollution yields a net benefit to society, since the social value of the increase in tax revenue due to the broadening of the pollution tax base more than offsets the environmental damage costs. Accordingly, this calls for subsidies rather than taxes on labor.

4.6.4   Optimal tax system with fixed labor tax

If the labor tax would be fixed exogenously, optimizing the tax system with respect to pollution and profit taxes yields the following solution for pollution taxes:

$$\theta_E = \frac{u_M \gamma_E}{\lambda} - \theta_L \frac{\epsilon_{LL}}{\epsilon_{LE}} \tag{4.28}$$

If we ignore environmental considerations ($u_M = 0$), the first term on the RHS of (4.28) would be zero. In that case, the second term on the RHS of (4.28) reveals that initial distortions in the tax system due to a positive labor tax (i.e. $T_L > 0$) call for pollution subsidies, rather than taxes. The reason is that the government has access to a lump-sum tax to raise sufficient revenues without distorting the allocation of inputs. Pollution subsidies should then be adopted to correct for distortions imposed by the labor tax.

If environmental considerations are taken into account, the optimal pollution tax in (4.28) contains two terms. First, pollution taxes would be used as instruments for environmental protection (see the first term on the RHS of (4.28)). The second term on the RHS of (4.28) shows that pollution subsidies should be used to alleviate initial distortions due to positive labor taxes.

4.7   Tax reform from labor to pollution taxes

This subsection explores the consequences of a rise in pollution taxes if the revenues are used to cut the labor tax. The reduced forms are presented in Table 4.8 (see appendix 4D for a derivation).[41]

4.7.1   Starting from a 100% profit tax

If the initial pollution tax is zero ($\theta_E = 0$), while profits are fully taxed away (i.e. $T_{\Pi} = 1$), the reduced forms in Table 4.8 reveal that a shift from labor towards pollution taxes leaves private income and employment unchanged. The reason is that the owners of the fixed factor receive zero profit income and thus cannot

---

[41] Bovenberg and de Mooij (1996) analyze the case of higher pollution taxes, recycled through lower profit taxes. Such a reform reduces private welfare, as the reform replaces non-distortionary profit taxes for distortionary taxes on pollution. Hence, the tax system moves away from its non-environmental optimum, thereby reducing blue welfare.

bear the burden of taxation. In fact, the government receives the return on fixed capital by imposing a 100% profit tax. Hence, it bears the full burden of lower profits. As a result, labor is the only source of private income that can bear the incidence of taxes. Just as in the benchmark model, the swap of pollution taxes for labor taxes thus boils down to substituting implicit for explicit taxes on labor. In the absence of initial pollution taxes, this leaves employment and private income unchanged.

If the initial pollution tax is positive ($\theta_E > 0$), an environmental tax reform from labor towards pollution produces a tax-burden effect, i.e. a positive private cost associated with pollution abatement. As in the benchmark model, the tax-burden effect is fully borne by wage incomes. This result is consistent with the optimal-tax results from (4.24) and (4.25), which reveal that the optimal tax structure between labor and pollution is equal to the benchmark model if $T_\Pi = 1$.

4.7.2   Starting from a less than 100% profit tax

If profits are not fully taxed away by the government (i.e. $T_\Pi < 1$), households receive not only labor income but also profit income. Hence, the incidence of both labor taxes and pollution taxes can be borne by two sources of income, rather than by labor income alone. Accordingly, a reshuffling in the tax mix may induce a shift in tax incidence between different income sources.

How the distribution of the tax burden over the two sources of income affects efficiency is illustrated by the expression for the optimal tax structure in (4.23). For given profit taxes, this expression reveals that it is optimal to levy the highest tax rate on the input that features the lowest supply and demand elasticities. If the initial tax system deviates from (4.23), then the initial distribution of the tax burden over labor taxes and pollution taxes is inefficient. Hence, by redistributing the tax burden across labor and pollution, the reform affects the efficiency of the tax system as a revenue-raising device, thereby potentially producing a blue dividend.

The second and third rows of Table 4.8 show under which circumstances a reform from labor towards pollution taxes makes the tax system more efficient from both an environmental and a revenue-raising perspective. It reveals that pollution drops if $\epsilon_{EE} - \epsilon_{LE} > 0$.[42] In that case, there is a green dividend. The effect on blue welfare depends on two effects (see the third row of Table 4.8). First, it depends on the tax-burden effect represented by the first term of the reduced-form equation. The intuition behind this term is that, as far as taxes bear on labor, it is more efficient from a revenue-raising perspective to adopt direct labor taxes, rather than pollution taxes (see above).

Table 4.8: Reduced-form coefficients for the pollution tax in the model with a fixed factor in which the labor tax is endogenous

	$\tilde{T}_L$
$\Delta_F \tilde{L}$	$\eta_{LL}^{*} (1 - T_\Pi)(\epsilon_{LL} - \epsilon_{EL}) \dfrac{w_E}{w_L}$ $- \eta_{LL} \theta_E \dfrac{w_E}{w_L} \epsilon_D$
$\Delta_F \tilde{E}$	$- \eta_{LL}^{*} (1 - T_\Pi)(\epsilon_{EE} - \epsilon_{LE})$ $- [1 - \theta_L - \eta_{LL} \theta_L] \epsilon_D$
$\Delta_F [(1 - \theta_L) w_L \tilde{W}$ $+ w_\Pi (\tilde{\Pi} - \tilde{T}_\Pi)]$	$- (1 - \theta_L) \theta_E w_E \epsilon_D$ $+ \eta_{LL}^{*} (1 - T_\Pi)[\theta_L(\epsilon_{LL} - \epsilon_{EL}) - \theta_E(\epsilon_{EE} - \epsilon_{LE})] w_E$

$$\Delta_F = \eta_{LL}^{*} (1 - T_\Pi) - \eta_{LL}[\theta_L \epsilon_{LL} + \theta_E \epsilon_{LE}] + (1 - \theta_L)\epsilon_{LL}$$

---

[42] The term $\epsilon_{EE} - \epsilon_{LE}$ is typically positive, thus indicating a fall in pollution (see Table 4.7). This term may be negative if polluting inputs are separable in production and, compared to labor, polluting inputs are a much better substitute for fixed capital. We rule out this possibility, however, and assume that own-price effects dominate cross-price effects.

The second term in the reduced-form equation for private welfare denotes the efficiency with which the government taxes away profit incomes. This term contains two components: first, an effect induced by lower labor taxes and, second, an effect induced by higher pollution taxes. The decline in the labor tax rate reduces wage costs. This stimulates labor demand if $\epsilon_{LL} - \epsilon_{EL} > 0$.[43] The rise in labor demand is not a sufficient condition to induce an actual increase in employment, however (see the first row of Table 4.8). In particular, in order to maintain the labor market equilibrium, the market wage rate needs to rise so that labor supply is encouraged. The higher wage rate involves a reduction in profit incomes. Hence, an environmental tax reform changes the distribution of incomes in favor of workers and at the expense of capital owners. We call this the *tax-shifting effect*.[44] The expansion of employment associated with the tax-shifting effect broadens the base of the labor tax, thereby increasing tax revenues endogenously if $\theta_L > 0$. Tax shifting from labor towards profits thus makes the tax system more efficient as a revenue-raising device.[45]

The second term between square brackets in the reduced-form equation for private welfare involves the impact of higher pollution taxes. These taxes reduce the demand for polluting inputs if $\epsilon_{EE} - \epsilon_{LE} > 0$, thereby eroding the base of the pollution tax if $\theta_E > 0$. Accordingly, they reduce private incomes through a tax-burden effect on profit incomes. Intuitively, pollution taxes are inefficient

---

[43] Table 4.7 reveals that $\epsilon_{LL}-\epsilon_{EL}$ is always positive if pollution or the fixed factor are separable. However, this term may be negative if labor is separable. In particular, it is negative if producers find it much easier to substitute pollution than labor for the fixed factor. In this section, we rule out this possibility and assume that $\epsilon_{LL} - \epsilon_{EL} > 0$.

[44] Note that the interpretation of this effect is somewhat different from that in section 4.3.

[45] The compensated, rather than the uncompensated, elasticity of labor supply measures the expansion in employment associated with the tax-shifting effect (see the first element in the reduced form for employment in Table 4.8). The reason is that the negative income effect on labor supply is offset by the fall in profits. Hence, only substitution effects matter. In case of the tax-burden effect, however, the uncompensated elasticity of labor supply measures the employment effect. This is because the reduction in after-tax wages associated with the tax-burden effect leaves profits unchanged (see the second element in the reduced form for employment in Table 4.8).

instruments to tax away profit incomes if initial pollution taxes are high and the demand for polluting inputs is elastic.

By expanding employment and reducing the demand for polluting inputs, a swap of pollution taxes for labor taxes may boost non-environmental welfare only if the marginal welfare gain of additional employment dominates the marginal loss in non-environmental welfare (due to a drop in polluting inputs). In addition, this welfare gain (represented by a positive second term in the reduced form for private welfare) should dominate the tax-burden effect on labor (represented by the first term in the reduced form for private welfare). In that case, compared to polluting inputs, labor is overtaxed initially so that substituting pollution taxes for labor taxes shifts the tax system towards its non-environmental optimum. In particular, a strong double dividend is feasible if four conditions are met. First, the fixed factor should be able to bear a large part of the tax burden. This requires a small profit tax and a large share of the fixed factor in production (implying that ϵ_D is small). Second, initial labor taxes should be large compared to pollution taxes. This implies that a rise in employment is valuable, whereas a drop in pollution is cheap. Third, labor demand should be elastic relative to pollution demand (as was also illustrated by the optimal tax structure in (4.23)). Finally, labor supply should be sufficiently elastic so that the tax-shifting effect is sufficiently large.[46]

4.8   Numerical simulations

This section illustrates the findings of the two different versions of the model numerically. First, we consider the model in which all capital is perfectly mobile internationally. As the supply of capital is likely to be more elastic in the long

---

[46] Labor cannot bear the burden of taxation if it is supplied with infinite wage elasticity ($\eta_{LL} \to \infty$). Hence, in that case the effect on private income is solely determined by the second term between square brackets in the reduced form for private income. In particular, labor is overtaxed compared to pollution if this term is positive (see also Bovenberg and van der Ploeg (1998b)). In contrast, if labor is fixed in supply ($\eta_{LL} = \eta_{LL}^* = \eta_L = 0$), employment amounts to a fixed factor. In that case, the tax-shifting effect is zero and a double dividend cannot be obtained.

run than in the short run, these simulations can, roughly speaking, be interpreted as the long-run consequences of environmental tax reforms (although other aspects may differ between the long run and the short run as well). In the second model, all capital is immobile and supplied inelastically. These outcomes could be interpreted as the short-term consequences of environmental tax reforms.

4.8.1   Calibration

The calibration of the model requires empirical information about various shares, demand elasticities and other parameters. Most shares in the model are based on the calibration of the benchmark model and are presented in the first column of Table 4.9. Initial capital taxes are large compared to energy taxes, but are smaller than labor taxes. Indeed, the capital income tax on a before-tax basis is set at one-third, which is also the tax rate on fixed capital. The factor-demand elasticities are derived from a number of empirical studies (see e.g. Pindyk (1979), Berndt and Wood (1979), Griffin (1981), Berndt and Hesse (1986), Hesse and Tarkka (1986), Ilmakunnas and Törmä (1989)). In particular, the second column in Table 4.9 presents our parameterization for the demand elasticities based on this literature. These empirical estimates denote price elasticities conditional on the level of output. In order to find the elasticities used in our model, we need to transform these estimates into elasticities conditional on the fixed factor or labor, rather than on output. These transformed elasticities are presented in the third and fourth columns of Table 4.9. All other calibrated parameters are given in the fifth column of Table 4.9.

Given the uncertainty surrounding some price elasticities, we employ sensitivity analysis. In particular, the debate on the substitutability versus complementarity between capital and energy calls for sensitivity analysis on the cross-price elasticities of capital and energy demand. Furthermore, the sensitivity with respect to the cross-price elasticity between labor and energy is employed. This also illustrates the importance of the empirical foundation of the production structure for the double-dividend debate.

Table 4.9: Calibration of the models with capital

Shares	Parameters from empirical studies	Conditional on capital	Conditional on employment	Other elasticities
$w_L = 0.7$	$e_{LL} = 0.25$	$\epsilon_{LL} = 1.00$	$\epsilon_{EL}^{*} = -.46$	$\eta_L = .3$
$w_E = 0.1$	$e_{LE} = -0.04$	$\epsilon_{LE} = 0.07$	$\epsilon_{EE}^{*} = .58$	$\eta_{LL} = .2$
$w_K = 0.2$	$e_{LK} = -0.06$	$\epsilon_{EL} = 0.49$	$\epsilon_{EK}^{*} = -.04$	$\eta_{LL}^{*} = .42$
or $w_H = 0.2$				
$\theta_L = 0.5$	$e_{EL} = -0.28$	$\epsilon_{EE} = 0.61$	$\epsilon_{KL}^{*} = -.67$	
$\theta_K = 0.25$	$e_{EE} = 0.50$		$\epsilon_{KE}^{*} = -.02$	
$\theta_E = 0.1$	$e_{EK} = -0.1$		$\epsilon_{KK}^{*} = .71$	
$T_{II} = 0.33$	$e_{KL} = -0.42$			
	$e_{KE} = -0.05$			
	$e_{KK} = 0.65$			

4.8.2   Model simulations with mobile capital

The first two columns of Table 4.10 show the consequences of an increase in energy taxes that raises the producer price of energy by 10% in the model with mobile capital. The table reveals that energy demand falls by about 6%, which implies a green dividend. If the revenues from the energy tax are used to cut labor taxes, employment and private wage incomes fall by 0.07% and 0.13%, respectively. Hence, there is no double dividend in that case. Compared to labor, energy thus seems to be overtaxed from a non-environmental point of view in EU economies. This conclusion is consistent with the normal case in the theoretical model of section 4.3. If the revenues from the energy tax are used to cut capital taxes, rather than labor taxes, private income rises by 0.28%, whereas employment increases by 0.16%. These findings suggests that in the long run, compared to energy, capital is typically overtaxed in Western European countries. Hence, in the terminology of section 4.3, we can say that the tax-shifting effect dominates the tax-burden effect.

Table 4.10: Effects of an increase in energy taxes that raises after-taxes energy prices for firms by 10% in the model with mobile capital, recycled through lower labor taxes or capital taxes

	Base Run $e_{KE} = -.05\ e_{EK} = -.1$		K/E better substitutes $e_{KE} = -.25\ e_{EK} = -.5$		K/E poorer substitutes $e_{KE} = .15\ e_{EK} = .3$	
	Lower T_L	Lower T_K	Lower T_L	Lower T_K	Lower T_L	Lower T_K
Employment	-0.07	0.16	0.03	0.16	-0.17	0.07
Capital	-0.83	4.21	1.27	6.29	-2.93	2.12
Pollution	-6.53	-5.54	-6.43	-7.46	-6.63	3.63
Blue welfare	-0.13	0.28	0.05	0.43	-0.30	0.13

If capital and energy are better substitutes for each other than was shown in the base-run simulation, substituting energy taxes for ordinary taxes on labor and capital becomes more favorable from a non-environmental point of view. This is illustrated by the third and fourth columns of Table 4.10. In particular, we find that a shift from labor taxes towards energy taxes improves blue welfare in this case (blue welfare rises by 0.05%). This is because, with easier substitution between capital and energy, such a shift in the tax system induces a large substitution effect towards capital. Accordingly, capital demand rises by 1.27%. The broadening of the base of the capital tax allows for further reductions in the endogenous labor tax and, therefore, higher after-tax wage incomes.[47]

If capital and energy would be poorer substitutes than in the base-run simulation, substituting energy for labor taxes would be less favorable for blue

---

[47] This exceptional case corresponds to footnote 32 in the theoretical model of section 4.3.

welfare (see fifth and sixth columns of Table 4.10). However, the conclusion that capital is overtaxed compared to energy would still hold.

4.8.3   Model simulations with fixed capital

Table 4.11 presents the simulations of an increase in the energy tax that raises the after-tax energy price for producers by 10% in the model with immobile capital that is fixed in supply. We find that energy demand drops by around 5%. If the revenues would be used to reduce the labor tax, wages and employment would rise by 0.24% and 0.23%, respectively, while profits would fall by 0.49%. For a representative household, the increase in wages dominates the drop in profits so that, on balance, private welfare rises by 0.03%. Hence, the tax-shifting effect outweighs the tax-burden effect. In the short run, EU economies thus slightly overtax labor compared to energy, even if environmental considerations are ignored.

If labor and energy are much better substitutes than in the base-run simulation, a shift from labor towards energy taxes is even more favorable for blue welfare (see the second column of Table 4.11). Indeed, private incomes rise by 0.20%. This is because employment rises more substantially (+0.68%) than in the base-run simulation, as the substitution effect towards labor is larger. In addition, the fixed factor bears a larger part of the tax burden, resulting in a more substantial fall in profits of 0.74%. Consequently, the tax-shifting effect from profit incomes towards wages is more substantial.

A number of empirical estimates suggest that compared to capital, labor is a poorer substitute for energy. In particular, for nine European countries Hesse and Tarkka (1986) find a median substitution elasticity between energy and capital of 0.5, but a median elasticity of 0.2 between labor and energy. Hence, the results with easier substitution between E and L as presented in the second column of Table 4.11 may be unrealistic. The case with poorer substitution between energy and labor may be more relevant for most European countries. The third column in Table 4.11 reveals that this would be rather unfavorable for the double dividend in the short run. In particular, if labor and energy would be poorer substitutes than in the base-run simulation, blue welfare may even fall (-

0.03%), whereas the rise in employment vanishes. Hence, with poor substitution between labor and energy, labor does not seem to be overtaxed.

Table 4.11: Effects of an increase in energy taxes that raises after-tax energy prices for firms by 10% in the model with fixed capital, recycled through lower labor taxes

	Base Run $e_{EL}=-.28, e_{LE}=-.04$	E/L better substitutes $e_{EL}=-.49, e_{LE}=-.07$	E/L poorer substitutes $e_{EL}=-.07, e_{LE}=-.01$
Employment	0.23	0.68	0
Pollution	-4.87	-4.41	-4.61
Wage rate	0.24	0.82	0.04
Gross profits	-0.49	-0.74	-0.37
Blue welfare	0.03	0.20	0.03

4.9 Conclusions

This chapter shows that factors of production other than labor or polluting inputs -- such as clean capital or land -- are important for the non-environmental consequences of environmental tax reforms. Indeed, this chapter shows that incorporating capital creates the potential for a double dividend by inducing a so-called tax-shifting effect: an environmental tax reform may move the tax system closer towards its non-environmental optimum if the tax system is sub-optimal initially.

The first tax-shifting effect is illustrated in a model in which capital is perfectly mobile internationally. In that case, zero capital and pollution taxes are optimal from a non-environmental point of view, so that the tax system is sub-optimal if capital is taxed initially. By expanding capital demand and reducing the demand for polluting inputs, a swap of pollution taxes for capital taxes boosts non-environmental welfare if the marginal welfare gain of additional capital dominates the tax-burden effect, i.e. the marginal loss in non-environmental welfare due to a drop in polluting inputs. In that case, compared to polluting inputs, capital is overtaxed initially so that substituting pollution taxes for capital taxes shifts the tax system towards its non-environmental optimum. Our numerical simulations reveal that this is potentially important for EU economies.

The second tax-shifting effect is illustrated in a model in which capital shares in the burden of pollution taxes. In particular, the model includes immobile capital that is supplied inelastically. In that case, the initial distribution of taxes over labor and pollution may be inefficient from a non-environmental point of view, as the incidence of taxes not only falls on labor but also on income from the fixed factor. Indeed, a swap of pollution taxes for labor taxes may shift the burden of taxation from labor towards the return on fixed capital, thereby improving non-environmental welfare. Accordingly, the incorporation of fixed capital increases the potential for a tax-shifting effect and thus for a double dividend.

Hence, whereas labor taxes are the most efficient instruments from a non-environmental point of view if capital is supplied with infinite elasticity, the optimal labor tax is zero if capital is in fixed supply. Similarly, the optimal capital tax is zero if capital is perfectly mobile across borders, while capital taxes are the most efficient instrument to raise public revenues if they are in fixed supply. This illustrates the role of capital mobility for the optimal tax structure between labor, capital and pollution.

Appendix 4A   Optimal taxes in the model with mobile capital

We assume that the production function, $F(L,K,E)$ features CRS. From (3.2) and (4.1), we can derive the factor-demand equations for polluting inputs and capital, conditional on the level of employment:

$$E = Le[1 + T_K, 1 + T_E] \quad ; \quad K = Lk[1 + T_K, 1 + T_E] \qquad (4A.1)$$

where we have normalized units so that the market prices for E and K are equal to unity. By substituting (4A.1) into (3.1), we can express the producer wage rate, $W^P = (1+T_L)W$, as an implicit function of the producer prices of polluting inputs and capital:

$$W^P = w[1 + T_E, 1 + T_K] \qquad (4A.2)$$

where we derive from the non-profit condition (4.2):

$$\frac{\partial w}{\partial (1 + T_E)} = -\frac{E}{L} \quad ; \quad \frac{\partial w}{\partial (1 + T_K)} = -\frac{K}{L} \qquad (4A.3)$$

The optimal tax problem amounts to optimizing household utility with respect to the tax rates on labor (T_L), capital (T_K) and polluting inputs (T_E), given the optimizing behavior of private firms and households and the government budget constraint (4.3). We do not optimize with respect to the amount of public goods provision, so that the government's revenue requirement is constant, i.e. $dG = 0$. Accordingly, we have the following optimization problem:

$$\underset{T_L, T_K, T_E}{MAX} \quad \mathcal{L} = V(1, W, G, M) + \mu[T_L WL + T_K K + T_E E] \qquad (4A.4)$$

where μ denotes the shadow price of public funds and V is the indirect utility function:

$$V(1, W, G, M) = \underset{}{max} \quad u(C, V, G, M)$$
$$s.t. \quad WL = C \qquad (4A.5)$$

The first-order conditions of this optimization problem are:

$$\frac{\partial \mathcal{L}}{\partial T_L} = \frac{\partial V}{\partial W}\frac{\partial W}{\partial T_L} + \frac{\partial V}{\partial M}\frac{\partial M}{\partial E}\frac{\partial E}{\partial T_L} +$$

$$\mu\left[WL + T_L L\frac{\partial W}{\partial T_L} + T_L W\frac{\partial L}{\partial T_L} + T_K\frac{\partial K}{\partial T_L} + T_E\frac{\partial E}{\partial T_L}\right] = 0 \qquad (4A.6)$$

$$\frac{\partial \mathcal{L}}{\partial T_K} = \frac{\partial V}{\partial W}\frac{\partial W}{\partial T_K} + \frac{\partial V}{\partial M}\frac{\partial M}{\partial E}\frac{\partial E}{\partial T_K} +$$

$$\mu\left[T_L L\frac{\partial W}{\partial T_K} + T_L W\frac{\partial L}{\partial T_K} + K + T_K\frac{\partial K}{\partial T_K} + T_E\frac{\partial E}{\partial T_K}\right] = 0 \qquad (4A.7)$$

$$\frac{\partial \mathcal{L}}{\partial T_E} = \frac{\partial V}{\partial W}\frac{\partial W}{\partial T_E} + \frac{\partial V}{\partial M}\frac{\partial M}{\partial E}\frac{\partial E}{\partial T_E} +$$

$$\mu\left[T_L L\frac{\partial W}{\partial T_E} + T_L W\frac{\partial L}{\partial T_E} + T_K\frac{\partial K}{\partial T_E} + E + T_E\frac{\partial E}{\partial T_E}\right] = 0 \qquad (4A.8)$$

We write these equations in different notation. First, using the fact that $W=W^P/(1+T_L)$, we derive:

$$\frac{\partial W}{\partial T_L} = -\frac{W}{1+T_L} \quad ; \quad \frac{\partial W}{\partial T_K} = -\frac{K}{(1+T_L)L} \quad ; \quad \frac{\partial W}{\partial T_E} = -\frac{E}{(1+T_L)L} \qquad (4A.9)$$

Second, from Roy's identity, we know:

$$\frac{\partial V}{\partial W} = \lambda L \qquad (4A.10)$$

Finally, define the following elasticities:

$$\frac{\partial k}{\partial T_K}\frac{1+T_K}{k} = -\Gamma_{KK} \quad ; \quad \frac{\partial k}{\partial T_E}\frac{1+T_E}{k} = -\Gamma_{KE}$$

$$\frac{\partial e}{\partial T_E}\frac{1+T_E}{e} = -\Gamma_{EE} \quad ; \quad \frac{\partial e}{\partial T_K}\frac{1+T_K}{e} = -\Gamma_{EK} \qquad (4A.11)$$

The elasticities in (4A.11) represent changes in the polluting input/labor ratio (E/L) and the capital/labor ratio (K/L) as a result of price changes. By using

(4A.1) and (4A.11), substituting (4A.9) and (4A.10) and dividing through by $\frac{\partial W}{\partial T_i}$ for $i = L,K,E$, respectively, we rewrite (4A.6) - (4A.8) as follows:

$$[\,\theta_L + \theta_K\frac{w_K}{w_L} + (\theta_E - \frac{1}{\eta}\frac{u_M\gamma_E}{\lambda})\frac{w_E}{w_L}\,]\,\eta_{LL} = (1-\theta_L)(1 - \frac{1}{\eta}) \qquad (4A.12)$$

$$[\,\theta_L + \theta_K\frac{w_K}{w_L} + (\theta_E - \frac{1}{\eta}\frac{u_M\gamma_E}{\lambda})\frac{w_E}{w_L}\,]\,\eta_{LL} +$$
$$\theta_K\Gamma_{KK} + (\theta_E - \frac{1}{\eta}\frac{u_M\gamma_E}{\lambda})\Gamma_{KE} = (1-\theta_L)(1 - \frac{1}{\eta}) \qquad (4A.13)$$

$$[\,\theta_L + \theta_K\frac{w_K}{w_L} + (\theta_E - \frac{1}{\eta}\frac{u_M\gamma_E}{\lambda})\frac{w_E}{w_L}\,]\,\eta_{LL} +$$
$$\theta_K\Gamma_{EK} + (\theta_E - \frac{1}{\eta}\frac{u_M\gamma_E}{\lambda})\Gamma_{FF} = (1-\theta_L)(1 \quad \frac{1}{\eta}) \qquad (4A.14)$$

where $\eta = \mu/\lambda$ denotes the marginal cost of public funds and we used the fact that $w_E\Gamma_{EK} = w_K\Gamma_{KE}$. From the first-order conditions in (4A.12) - (4A.14), we can derive the optimal taxes presented in section 4.2. In particular, if the government cannot optimize with respect to either of the three taxes, the corresponding first-order condition does not necessarily hold, so that the remaining two conditions determine the optimal tax structure.

Appendix 4B   Tax reform in the model with mobile capital

To solve the model, we substitute equation (T4.1) from Table 4.1, into (T4.10) to eliminate $\tilde{Y}$:

$$w_L\tilde{L} = w_C\tilde{C} - w_E\theta_E\tilde{E} - w_K\theta_K\tilde{K} \qquad (4B.1)$$

We will express the endogenous variables on the RHS of (4B.1) in terms of $\tilde{L}$ and $\tilde{W}$. To find $\tilde{C}$, we know from (T4.5) (in the absence of polluting consumption):

$$w_C \tilde{C} = w_L (1 - \theta_L)(\tilde{W} + \tilde{L}) \qquad (4B.2)$$

To find $\tilde{E}$ and $\tilde{K}$ in terms of $\tilde{L}$ and $\tilde{W}$, we solve two versions of the model. First, we derive the reduced forms if the revenues from higher pollution taxes are used to reduce T_L endogenously. Subsequently, we do the same for the case that the revenues are used to cut T_K.

Endogenous labor tax

To eliminate the wage costs, $\tilde{W} + \tilde{T}_L$, from (T4.2) and (T4.3), we substitute (T4.4) and find for polluting input demand and capital demand:

$$\tilde{E} = \tilde{L} - \Gamma_{EE} \tilde{T}_E - \Gamma_{EK} \tilde{T}_K \qquad (4B.3)$$

$$\tilde{K} = \tilde{L} - \Gamma_{KE} \tilde{T}_E - \Gamma_{KK} \tilde{T}_K \qquad (4B.4)$$

where

$$\Gamma_{EE} = \epsilon_{EE}^* - \frac{w_E}{w_L} \epsilon_{EL}^* \quad ; \quad \Gamma_{EK} = \epsilon_{EK}^* - \frac{w_K}{w_L} \epsilon_{EL}^*$$

$$\Gamma_{KE} = \epsilon_{KE}^* - \frac{w_E}{w_L} \epsilon_{KL}^* \quad ; \quad \Gamma_{KK} = \epsilon_{KK}^* - \frac{w_K}{w_L} \epsilon_{KL}^* \qquad (4B.5)$$

Assuming that $\tilde{T}_K = 0$ and substituting (4B.2), (4B.3) and (4B.4) into (4B.1), we find the following relationship between $\tilde{L}$, $\tilde{W}$, and $\tilde{T}_E$:

$$[\theta_L w_L + \theta_E w_E + \theta_K w_K] \tilde{L} - (1 - \theta_L) w_L \tilde{W} = [\theta_E \Gamma_{EE} + \theta_K \Gamma_{EK}] w_E \tilde{T}_E \qquad (4B.6)$$

where we used $w_K \Gamma_{KE} = w_E \Gamma_{EK}$. Together with the relationship for labor supply in (T4.6) this forms a set of two equations and two endogenous variables $\tilde{L}$ and $\tilde{W}$. In matrix notation, we write them as follows:

$$\begin{pmatrix} \theta_L w_L + \theta_E w_E + \theta_K w_K & -(1-\theta_L)w_L \\ 1 & -\eta_{LL} \end{pmatrix} \begin{pmatrix} \tilde{L} \\ \tilde{W} \end{pmatrix} =$$

$$\begin{pmatrix} (\theta_E \Gamma_{EE} + \theta_K \Gamma_{EK})w_E \\ 0 \end{pmatrix} \tilde{T}_E \tag{4B.7}$$

By inverting the matrix on the LHS of (4B.7), we derive:

$$\Delta_L \begin{pmatrix} \tilde{L} \\ \tilde{W} \end{pmatrix} = \begin{pmatrix} -\eta_{LL} & (1-\theta_L)w_L \\ -1 & \theta_L w_L + \theta_E w_E + \theta_K w_K \end{pmatrix} \begin{pmatrix} (\theta_E \Gamma_{EE} + \theta_K \Gamma_{EK})w_E) w_K \\ 0 \end{pmatrix} \tilde{T}_E \tag{4B.8}$$

where

$$\Delta_L = (1-\theta_L)w_L - \eta_{LL}[\theta_L w_L + \theta_E w_E + \theta_K w_K] > 0 \tag{4B.9}$$

From (4B.8), we derive the reduced forms for employment and wages:

$$\Delta_L \tilde{L} = -\eta_{LL}[\theta_E \Gamma_{EE} + \theta_K \Gamma_{EK}]w_E \tilde{T}_E \tag{4B.10}$$

$$\Delta_L \tilde{W} = -[\theta_E \Gamma_{EE} + \theta_K \Gamma_{EK}]w_E \tilde{T}_E \tag{4B.11}$$

By substitution of (4B.10) into (4B.3) and (4B.4), we derive the reduced forms for polluting input demand and capital demand. The reduced form for output is found by substituting (4B.10) and the reduced forms for pollution and capital into (T4.1).

Endogenous capital income tax
If the revenues from environmental taxes are used to cut the capital income tax, rather than the labor income tax, we substitute (T4.4) into (T4.2) and (T4.3) to

eliminate the capital income tax. Thus, we find for polluting input and capital demand:

$$\tilde{E} = \tilde{L} - (\Gamma_{EE} - \Gamma_{KE})\tilde{T}_E + \frac{w_L}{w_K}\Gamma_{EK}(\tilde{W} + \tilde{T}_L) \qquad (4B.12)$$

$$\tilde{K} = \tilde{L} + \frac{w_E}{w_K}(\Gamma_{KK} - \Gamma_{EK})\tilde{T}_E + \frac{w_L}{w_K}\Gamma_{KK}(\tilde{W} + \tilde{T}_L) \qquad (4B.13)$$

Substituting (4B.12) and (4B.13) as well as (4B.2) into (4B.1), we find a relationship between $\tilde{L}, \tilde{W}$ and $\tilde{T}_E$ (assuming $\tilde{T}_L - 0$):

$$[\theta_L w_L + \theta_E w_E + \theta_K w_K]\tilde{L} - [(1 - \theta_L)w_L - w_L(\theta_E\Gamma_{KE} + \theta_K\Gamma_{KK})]\tilde{W}$$
$$\hspace{6cm} (4B.14)$$
$$= [\theta_E(\Gamma_{EE} - \Gamma_{KE}) - \theta_K(\Gamma_{KK} - \Gamma_{EK})]w_E\tilde{T}_E$$

Combining (4B.14) and (T4.6) and applying the method of matrix inversion, we find the following reduced forms for employment and wages:

$$\Delta_K\tilde{L} = -\eta_{LL}[\theta_E(\Gamma_{EE} - \Gamma_{KE}) - \theta_K(\Gamma_{KK} - \Gamma_{EK})]w_E\tilde{T}_E \qquad (4B.15)$$

$$\Delta_K\tilde{W} = -[\theta_E(\Gamma_{EE} - \Gamma_{KE}) - \theta_K(\Gamma_{KK} - \Gamma_{EK})]w_E\tilde{T}_E \qquad (4B.16)$$

where

$$\Delta_K = (1 - \theta_L)w_L - \theta_E w_L\Gamma_{KE} - 0_K w_L\Gamma_{KK} - \eta_{LL}[\theta_L w_L + \theta_E w_E + \theta_K w_K] \qquad (4B.17)$$

By substituting (4B.15) into (4B.12) and (4B.13), we arrive at the reduced forms for polluting input demand and capital demand. The solution for output is found by substituting the reduced forms for employment, polluting inputs and capital into (T4.1).

Appendix 4C   Optimal taxes in the model with fixed capital

To solve the optimal tax problem, we start by deriving the factor-demand equations from (3.1) and (3.2), conditional on the fixed factor:

$$L = Hl[(1 + T_L)W, 1 + T_E] \quad ; \quad E = He[(1 + T_L)W, 1 + T_E] \qquad (4C.1)$$

By substituting $F(L,E,H)$ into (4.17), we can express profits as an implicit function of the producer prices of polluting inputs and labor:

$$\Pi = \pi \left[(1 + T_L) W, \, 1 + T_E \right] \tag{4C.2}$$

where we derive from the profit equation (4.17):

$$\frac{\partial \pi}{\partial (1 + T_E)} = -E \quad ; \quad \frac{\partial \pi}{\partial (1 + T_L) W} = -L \tag{4C.3}$$

The optimal tax problem is set up by optimizing household utility with respect to the tax rates on labor (T_L), polluting inputs (T_E) and profits (T_Π), i.e.:

$$\underset{T_L, T_E, T_\Pi}{MAX} \quad \mathcal{L} = V(1, W, G, M) + \mu \left[T_L WL + T_E E + T_\Pi \Pi \right] \tag{4C.4}$$

where μ denotes the shadow price of public funds, V is the indirect utility function, and the household budget constraint is given by (4.18). The first-order conditions of the optimal tax problem are:

$$\frac{\partial \mathcal{L}}{\partial T_L} = \frac{\partial V}{\partial (1 - T_\Pi) \Pi} (1 - T_\Pi) \frac{\partial \Pi}{\partial T_L} + \frac{\partial V}{\partial W} \frac{\partial W}{\partial T_L} + \frac{\partial V}{\partial M} \frac{\partial M}{\partial E} \frac{\partial E}{\partial T_L}$$

$$+ \mu \left[WL + T_L L \frac{\partial W}{\partial T_L} + T_L W \frac{\partial L}{\partial T_L} + T_E \frac{\partial E}{\partial T_L} + T_\Pi \frac{\partial \Pi}{\partial T_L} \right] = 0 \tag{4C.5}$$

$$\frac{\partial \mathcal{L}}{\partial T_E} = \frac{\partial V}{\partial (1 - T_\Pi) \Pi} (1 - T_\Pi) \frac{\partial \Pi}{\partial T_E} + \frac{\partial V}{\partial W} \frac{\partial W}{\partial T_E} + \frac{\partial V}{\partial M} \frac{\partial M}{\partial E} \frac{\partial E}{\partial T_E}$$

$$+ \mu \left[T_L L \frac{\partial W}{\partial T_E} + T_L W \frac{\partial L}{\partial T_E} + T_E \frac{\partial E}{\partial T_E} + E + T_\Pi \frac{\partial \Pi}{\partial T_L} \right] = 0 \tag{4C.6}$$

$$\frac{\partial \mathcal{L}}{\partial T_E} = \frac{\partial V}{\partial (1 - T_\Pi) \Pi} (1 - T_\Pi) \frac{\partial \Pi}{\partial T_\Pi} + \frac{\partial V}{\partial W} \frac{\partial W}{\partial T_\Pi} + \frac{\partial V}{\partial M} \frac{\partial M}{\partial E} \frac{\partial E}{\partial T_\Pi}$$

$$+ \mu \left[T_L L \frac{\partial W}{\partial T_\Pi} + T_L W \frac{\partial L}{\partial T_\Pi} + T_E \frac{\partial E}{\partial T_\Pi} + T_\Pi \frac{\partial \Pi}{\partial T_\Pi} + \Pi \right] = 0 \tag{4C.7}$$

Define the following elasticities:

$$\frac{\partial e}{\partial 1 + T_E} \frac{1 + T_E}{e} = -\epsilon_{EE} \quad ; \quad \frac{\partial e}{\partial (1 + T_L) W} \frac{(1 + T_L) W}{e} = -\epsilon_{EL}$$

$$\frac{\partial l}{\partial 1 + T_E} \frac{1 + T_E}{l} = -\epsilon_{LE} \quad ; \quad \frac{\partial l}{\partial (1 + T_L) W} \frac{(1 + T_E) W}{l} = -\epsilon_{LL} \qquad (4C.8)$$

$$\frac{\partial L_S}{\partial W} \frac{W}{L_S} = \eta_{LL} \quad ; \quad \frac{\partial L_S}{\partial (1 - T_\Pi) \Pi} \frac{C}{L_S} = -\eta_L$$

Labor demand is implicitly determined by (4C.1), while labor supply can be derived from (3.5) as a function of the relative prices of leisure and consumption and the after-tax profits:

$$L_S = l_s[W, (1 - T_\Pi)\Pi] \qquad (4C.9)$$

A change in either tax rate has an equivalent effect on both labor supply and labor demand, as wages equilibrate the labor market (i.e. $\dfrac{dL_S}{dT_i} = \dfrac{dL_D}{dT_i}$).

Accordingly, we derive the effect of various tax changes on the wage rate:

$$\frac{\partial W}{\partial T_L} \frac{1 + T_L}{W} = -\frac{\epsilon_{LL} + (1 - T_\Pi)\dfrac{w_L}{w_C}\eta_L}{\epsilon_{LL} + \eta_{LL} + (1 - T_\Pi)\dfrac{w_L}{w_C}\eta_L} \qquad (4C.10)$$

$$\frac{\partial W}{\partial T_E} \frac{1 + T_E}{W} = -\frac{\epsilon_{LE} + (1 - T_\Pi)\dfrac{w_E}{w_C}\eta_L}{\epsilon_{LL} + \eta_{LL} + (1 - T_\Pi)\dfrac{w_L}{w_C}\eta_L} \qquad (4C.11)$$

$$\frac{\partial W}{\partial T_{\Pi}} \frac{1 + T_{\Pi}}{W} = - \frac{(1 - T_{\Pi}) \frac{w_{\Pi}}{w_C} \eta_L}{\epsilon_{LL} + \eta_{LL} + (1 - T_{\Pi}) \frac{w_L}{w_C} \eta_L} \qquad (4C.12)$$

By using the definitions in (4C.8), Roy's identity from (4A.10), and relations (4C.10) - (4C.12), we rewrite the first-order conditions in (4C.5) - (4C.7) as follows:

$$\theta_L \epsilon_{LL} + (\theta_E - \frac{1}{\eta} \frac{u_M \gamma_E}{\lambda}) \epsilon_{LE} = (1 - \frac{1}{\eta}) \frac{1}{\eta_{LL}} [(1 - T_{\Pi}) \eta_{LL}^* + \frac{\epsilon_{LL}}{1 + T_L}] \qquad (4C.13)$$

$$\theta_L \epsilon_{EL} + (\theta_E - \frac{1}{\eta} \frac{u_M \gamma_E}{\lambda}) [\epsilon_{EE} + \frac{\epsilon_D}{\eta_{LL}}]$$

$$- (1 + T_L)(1 - T_{\Pi}) \frac{\eta_{LL}^* - \eta_{LL}}{\eta_{LL}} [\theta_L (\epsilon_{LL} - \epsilon_{EL}) - (\theta_E - \frac{1}{\eta} \frac{u_M \gamma_E}{\lambda})(\epsilon_{EE} - \epsilon_{LE})] \qquad (4C.14)$$

$$= (1 - \frac{1}{\eta}) \frac{1}{\eta_{LL}} [(1 - T_{\Pi})(\eta_{LL}^* + \epsilon_{LL} - \epsilon_{EL}) + \frac{\epsilon_{EL}}{1 + T_L}]$$

$$\theta_L \epsilon_{LL} + (\theta_E - \frac{1}{\eta} \frac{u_M \gamma_E}{\lambda}) \epsilon_{LE} = - (1 - \frac{1}{\eta}) \frac{\eta_{LL}^* + \epsilon_{LL}}{\eta_{LL}^* - \eta_{LL}} \frac{1}{1 + T_L} \qquad (4C.15)$$

From (4C.13) - (4C.15), the solutions for the optimal tax rates, presented in section 4.4, can be derived.

Appendix 4D   Tax reform in the model with fixed capital

This appendix solves the model with fixed capital if the labor tax is endogenously adjusted to balance the government budget. The profit tax is assumed to be fixed. To solve the model, we derive three equations in terms of the three endogenous variables $\tilde{L}$, $\tilde{W}$ and $\tilde{T}_L$ and the exogenous tax rate, $\tilde{T}_E$.

First, we subtract (T4.20) from (T4.11) from Table 4.6 to eliminate $\tilde{Y}$:

$$w_L \tilde{L} = w_C \tilde{C} - w_E \theta_E \tilde{E} \tag{4D.1}$$

By substituting (T4.12) and (T4.13) into (4D.1) to eliminate $\tilde{E}$ and $\tilde{C}$, respectively, we find:

$$\theta_L w_L \tilde{L} = w_L(1 - \theta_L)\tilde{W} + w_\Pi \tilde{\Pi} + w_E \theta_E [\epsilon_{EL}(\tilde{W} + \tilde{T}_L) + \epsilon_{EE}\tilde{T}_E] \tag{4D.2}$$

Profits can be eliminated from (4D.2) by substituting (T4.14):

$$\theta_L w_L \tilde{L} = w_L(1 - \theta_L)\tilde{W} - (1 - T_\Pi)[w_L(\tilde{W} + \tilde{T}_L) + w_E \tilde{T}_E] + \\ w_E \theta_E [\epsilon_{EL}(\tilde{W} + \tilde{T}_L) + \epsilon_{EE}\tilde{T}_E] \tag{4D.3}$$

Second, by substituting (T4.14) into (T4.16), we find for labor supply:

$$\tilde{L}_S = \eta_{LL}\tilde{W} + \frac{\eta_L}{w_C}(1 - T_\Pi)[w_L(\tilde{W} + \tilde{T}_L) + w_E \tilde{T}_E] \tag{4D.4}$$

Relations (T4.13), (4D.3) and (4D.4) form three equations in terms of $\tilde{L}$, $\tilde{W}$, $\tilde{T}_L$, and $\tilde{T}_E$. They can be written in matrix notation as:

$$\begin{pmatrix} 1 & \epsilon_{LL} & \epsilon_{LL} \\ 1 & -[\eta_{LL} + \eta_L(1 - T_\Pi)\dfrac{w_L}{w_C}] & -\eta_L(1 - T_\Pi)\dfrac{w_L}{w_C} \\ \theta_L w_L & -w_L[(1 - \theta_L) - (1 - T_\Pi) + \theta_E \epsilon_{LE}] & w_L[(1 - T_\Pi) - \theta_E \epsilon_{LE}] \end{pmatrix} \begin{pmatrix} \tilde{L} \\ \tilde{W} \\ \tilde{T}_L \end{pmatrix}$$

$$= \begin{pmatrix} -\epsilon_{LE} \\ \eta_L(1 - T_\Pi)\dfrac{w_E}{w_C} \\ -(1 - T_\Pi)w_E + \theta_E w_E \epsilon_{EE} \end{pmatrix} \tilde{T}_E \tag{4D.5}$$

Applying Cramer's Rule to (4D.5), we find the following solutions for employment, the wage rate and the labor tax:

$$\Delta_F \tilde{L} = \eta_{LL}^* (1 - T_\Pi)(\epsilon_{LL} - \epsilon_{EL}) w_E \tilde{T}_E - \eta_{LL} \theta_E w_E \epsilon_D \tilde{T}_E \qquad (4D.6)$$

$$\Delta_F \tilde{W} = [(1 - T_\Pi)(\epsilon_{LL} - \epsilon_{EL}) - \theta_E \epsilon_D] w_E \tilde{T}_E +$$
$$(1 - T_\Pi)\frac{w_L}{w_C} \eta_L [\theta_L(\epsilon_{LL} - \epsilon_{EL}) - \theta_E(\epsilon_{EE} - \epsilon_{LE})] w_E \tilde{T}_E \qquad (4D.7)$$

$$\Delta_F \tilde{T}_L = -\eta_{LL}[(1 - T_\Pi)w_E - \theta_L w_L \epsilon_{LE} - \theta_E w_E \epsilon_{EE}]\tilde{T}_E -$$
$$[(1 - T_\Pi)(\epsilon_{LL} - \epsilon_{EL}) - \theta_E \epsilon_D] w_E \tilde{T}_E -$$
$$(1 - T_\Pi)\frac{w_L}{w_C} \eta_L w_E [\theta_L(\epsilon_{LL} - \epsilon_{EL}) - \theta_E(\epsilon_{EE} - \epsilon_{LE})]\tilde{T}_E -$$
$$(1 - \theta_L)w_L [\epsilon_{LE} + (1 - T_\Pi)\eta_L \frac{w_E}{w_C}]\tilde{T}_E \qquad (4D.8)$$

where

$$\Delta_F = \eta_{LL}[(1 - T_\Pi)w_L - \theta_L w_L \epsilon_{LL} - \theta_E w_E \epsilon_{EL}] + (1 - \theta_L)w_L[\epsilon_{LL} + (1 - T_\Pi)\eta_L \frac{w_L}{w_C}] \qquad (4D.9)$$

$\epsilon_D = \epsilon_{LL}\epsilon_{EE} - \epsilon_{LE}\epsilon_{EL}$ and $\eta_{LL}^* = \eta_{LL} + \eta_L \dfrac{(1 - \theta_L)w_L}{w_Q}$ is the compensated

elasticity of labor supply. The determinant in (4D.9) should be positive for the equilibrium to be stable. The reduced forms for pollution and profits are found by substituting (4D.7) and (4D.8) into (T4.12) and (T4.14), respectively.

5 Environmental taxes as trade-policy instruments

This chapter explores how terms-of-trade effects influence the double-dividend analysis. The benchmark model assumes that prices of tradable commodities are fixed and determined exogenously on world markets. This assumption is appropriate for a small open economy, but cannot be used for economies that exert market power on international markets, especially coalitions of countries. To obtain more insight into the double-dividend analysis for countries that possess market power, this chapter develops a simple three-country model. The home economy -- which features preexisting distortions associated with labor taxes and environmental externalities -- may change world-market prices of polluting inputs and intermediate inputs. Within this framework, I explore the welfare effects of a unilateral environmental tax reform in the home country. Furthermore, this chapter derives the optimal pollution tax in the home country, both from a national and from a global point of view.

The theoretical literature on the double dividend has not paid much attention to terms-of-trade effects. In their overview articles, Goulder (1995a) and Bovenberg (1997) mention the potential for a double dividend through terms-of-trade gains if, for instance, the US or a group of countries that import fossil fuels coordinate their actions. Neither of these authors formally examine this issue, however. Killinger (1995) explores the double dividend in a two-country model in which the home country exerts market power on the world capital market. He finds that pollution taxes may partly shift the tax burden to foreign suppliers of capital, since these taxes tend to reduce the world-wide rate of return to capital. More recently, Smulders and Sen (1998) explore the welfare effects of an increase in pollution taxes, recycled through lower tariffs on an imported commodity. This might be especially relevant for developing countries, where governments raise substantial revenues through tariffs.[48]

In contrast to the theoretical literature, terms-of-trade effects have received a lot of attention in empirical studies on the cost of coordinated policies to reduce CO_2 emissions. Indeed, these studies generally suggest that terms-of-trade effects are important for the economic impact of coordinated CO_2 policies (see e.g. Hoeller et al., 1991; Hoeller et al., 1992; Whalley and Wigle, 1991). This chapter may help to understand these results from empirical models by demonstrating the importance of terms-of-trade effects in an analytical framework. Accordingly, it derives the crucial parameters for a double dividend through the impact on the terms of trade.

This chapter is closely related to the literature on optimal tariffs. This literature reveals that a large country that possesses market power on an international market may raise its welfare by imposing a tariff. In particular, although a tariff involves an efficiency cost by distorting the allocation of commodities or inputs, it may also improve the terms of trade of the country. If, over some range, the terms-of-trade gain exceeds the efficiency cost, the optimal tariff will be positive. Boadway et al. (1973) explore optimal tariffs in a second-best framework in which the government adopts distortionary taxes to finance public goods. They find that prior tax distortions magnify both the social costs and social benefits of tariffs, but leave the second-best optimal tariff unchanged, compared to its first-best level (see also Williams (1998a)). International free-trade agreements may preclude governments from using explicit tariffs or quotas to affect the terms of trade. In that case, pollution taxes may operate as implicit trade-policy instruments because they have similar effects on the terms of trade as explicit tariffs. Indeed, in the model of this chapter, pollution taxes are

---

[48] Most of the theory on environmental policy and trade focuses on the impact of environmental regulation on the competitive position of economies. On the one hand, a number of these studies suggest that environmental regulation will typically harm competitiveness (see e.g. McGuire, 1982). On the other hand, studies have found no clear-cut empirical evidence that environmental standards affect international trade flows (see e.g. Jaffe et al., 1995) or suggest that environmental policies can actually improve competitiveness (Barrett, 1994). This chapter does not explore the impact of pollution taxes on competitiveness, but concentrates instead on the impact of environmental policy on the terms of trade.

implicit import tariffs if the home country exerts monopsony power on the international market for polluting inputs. Accordingly, this chapter extends the analysis by Boadway et al. (1973) by investigating the welfare effects of pollution taxes -- in their role as implicit tariffs -- in a world with preexisting labor-market distortions and environmental externalities.

The rest of this chapter is organized as follows. Section 5.1 elaborates on the structure of the three-country model. Section 5.2 discusses environmental taxes in the home country if the price of polluting inputs is endogenous. Section 5.3 explores the implications of the reform if the home country exerts monopoly power on the output market. In section 5.4, I illustrate the importance of terms-of-trade effects with some numerical simulations. Finally, section 5.5 concludes.

5.1   The three-country model

This section develops a model describing three economies that are linked through trade in polluting inputs, (clean) intermediate inputs, and final goods. Each country specializes in the production of each of these commodities. In particular, country 1 (the home country) specializes in the production of (clean) intermediate goods, country 2 supplies polluting inputs, and country 3 produces final goods. The three economies are presented in Table 5.1. Country-specific variables are indexed by a subscript, while a superscript indicates demand or supply. Subsection 5.1.1 discusses the structure of the three-country model. Subsection 5.1.2 presents the linearized model for the home country, as far as it differs from the benchmark model. Finally, subsection 5.1.3 discusses the marginal excess burden to explore the welfare effects of public policies.

5.1.1   Structure of the model

Country 1: The home country
The economy of the home country is described by the benchmark model of chapter 3 and summarized in the first column of Table 5.1. In particular, the model describes an open economy, specialized in the production of intermediate

inputs (Y_I^S). These intermediate inputs are produced by a number of representative firms that combine domestic labor (L_I^D) with polluting inputs (E_I^D). These latter inputs are imported from country 2 and can be thought of as energy inputs. Domestic firms compete with each other on a perfectly competitive output market. The home economy exports her intermediate inputs to country 3, while it imports from country 3 the commodities that are consumed by domestic households (C_I^D) and by the government (G_I^D). The price of final goods is the numeraire. Country 1 differs from the economy described by the benchmark model in that consumers demand only clean consumption commodities (i.e. D is absent).

Table 5.1: The structure of the three-country model

Country 1 (home country)	Country 2 (OPEC)	Country 3 (rest of the world)
Production		Production
$Y_1^S = f_1[E_1^D, L_1^D]$		$C_3^S = f_3(L_3^D, E_3^D, Y_3^D)$
Input demand	Supply of polluting inputs	Input demand
$L_1^D = \alpha E_1^D\left(\dfrac{(1+T_L)W_1}{P_E+T_E}\right)^{-\sigma_{LE}}$	$E_2^S = P_E^{1/\gamma}$	$E_3^D = \psi L_3^D P_E^{-\epsilon_{EE}} P_Y^{-\epsilon_{EY}}$
		$Y_3^D = \kappa L_3^D P_E^{-\epsilon_{YE}} P_Y^{-\epsilon_{YY}}$
Non-profit condition		Non-profit condition
$P_Y Y_1^S = (1+T_L)W_1 L_1^D + (P_E+T_E)E_1^D$		$C_3^S = W_3 L_3^D + P_E E_3^D + P_Y Y_3^D$
Household utility	Household utility	Household utility
$U_1 = u_1[C_1^D, 1-L_1^S, G_1^D, M_1]$	$U_2 = u_2(C_2^D)$	$U_3 = u_3(C_3^D)$
Budget constraint	Budget constraint	Budget constraint
$W_1 L_1^S = C_1^D$	$P_E E_2^S = C_2^D$	$W_3 L_3^S = C_3^D$

Labor supply $L_1^S = W_1^{\eta_{LL}}$		
Balance of payments $P_Y Y_1^S = C_1^D + G_1^D + P_E E_1^D$	Balance of payments $P_E E_2^S = C_2^D$	Balance of payments $C_3^S = C_3^D + P_E E_3^D + P_Y Y_3^D$
Environmental quality $M_1 = m(E_1^D, E_3^D)$		
Labor-market equilibrium $L_1^D = L_1^S$		L_3^S is fixed $L_3^D = L_3^S$
Government budget $T_L W_1 L_1^D + T_E E_1^D = G_1^S$ $G_1^S = G_1^D$		

International market for intermediate inputs: $Y_1^S = Y_3^D$

International market for final goods: $C_3^S = C_1^D + G_1^D + C_2^D + C_3^D$

International market for polluting inputs: $E_2^S = E_1^D + E_3^D$

Country 2: The supplier of polluting inputs

The second column of Table 5.1 describes the economy of country 2. This country can be viewed as the group of OPEC countries that supply crude oil to the rest of the world. In particular, country 2 owns a resource stock that is unique in the world. By extracting resources from this stock, the country produces intermediate inputs (E_2^S) that pollute the environment when they are consumed. The supply of polluting inputs by country 2 is inversely related to the price of these inputs, P_E, where the parameter γ stands for the reciprocal of the supply elasticity. Country 2 is populated by representative households who own the firms in country 2. Households maximize utility (U_2), which contains the consumption of private commodities (C_2^D) as its only argument. The household budget available for consumption is determined by the revenues from the sale of polluting inputs.

Country 3: The rest of the world
Country 3 can be considered as the rest of the world. This economy is described by the third column of Table 5.1. Representative firms in country 3 produce final goods (C_3^S) by combining three inputs: domestic labor, which is in fixed supply (L_3^D), polluting inputs imported from country 2 (E_3^D), and intermediate inputs imported from country 1 (Y_3^D). The CES production function exhibits constant returns to scale with respect to the three inputs. Firms maximize profits and operate on a perfectly competitive world market for final goods. From the first-order conditions, one derives the two factor-demand equations: one for polluting inputs and one for intermediate inputs. The direct and indirect price elasticities in these factor-demand equations (ϵ_{ij}, i,j = E,Y) are defined relative to the level of employment in country 3 (see chapter 4 for an exposition of these elasticities for three separable production functions). Hence, households in country 3 receive only labor income, which is used for the consumption of final goods (C_3^D). Final goods are not only consumed by households in country 3, but are also exported to countries 1 and 2.

International markets
The three countries trade intermediate inputs, polluting inputs and final goods. Perfect competition ensures equilibrium on each of these international markets and yields equilibrium prices.[49] The price of final goods that are consumed in all countries is taken as the numeraire. The aim of this chapter is to explore how unilateral policies in the home country affect the terms of trade. In particular, the home country may be a non-negligible player on the market for polluting inputs. Hence, domestic policies may change the terms of trade by affecting the import price of polluting inputs (P_E). Furthermore, since country 1 as a whole is a

---

[49] Walras Law implies that one of those equilibrium conditions can be derived from the other expressions in the model.

monopolist in the production of intermediate inputs, domestic policies may affect the terms of trade through their impact on the price of intermediate inputs (P_Y).

The global environment
Households in the home country care about the quality of the environment (M_1) which deteriorates through the use of polluting inputs by both country 1 and country 3. Hence, the model allows pollution to cross international borders. To illustrate, energy consumption in all countries emits CO_2 emissions in the global atmosphere, thereby contributing to the global environmental problem of the greenhouse effect. The effect of national pollution on environmental quality can be larger than that of transboundary pollution. Indeed, the burning of fossil fuels causes not only CO_2 emissions, but also SO_2 emissions and local smog. Hence, the environmental externality in country 1 has both a global character and a local character. Households in countries 2 and 3 are not concerned with the quality of the environment. Hence, this chapter explores whether the home country is able to internalize the externality through unilateral policies alone.[50]

5.1.2 Linearization

This section discusses the linearized price equations of the model contained in Table 5.1. Furthermore, it presents the linearized model of the home economy insofar as it differs from the benchmark model. To simplify things, I assume that the cross-price elasticities for intermediate inputs and polluting inputs in country 3 are zero (i.e. $\epsilon_{YE} = \epsilon_{EY} = 0$).

The price of polluting inputs
Linearizing the expression for the supply of polluting inputs by country 2, one can write the price of polluting inputs as a function of aggregate demand for polluting inputs (which equals supply):

---

[50] The literature on policies to reduce greenhouse gas emissions has paid much attention to policy coordination: see e.g. Poterba (1993).

$$\tilde{P}_E = \gamma [s_1^E \tilde{E}_1^D + (1 - s_1^E) \tilde{E}_3^D] \qquad (5.1)$$

where $s_1^E = E_1^D/(E_1^D + E_3^D)$ stands for the share of polluting input demand by country 1 in terms of global demand for polluting inputs. According to (5.1), the impact of polluting input demand on the price of polluting inputs is measured by the parameter γ, which stands for the inverse-supply elasticity of polluting inputs. In particular, if γ is close to zero, polluting inputs are supplied elastically, so that the price of polluting inputs will hardly respond to demand shocks. However, if γ is large, polluting input supply is rather inelastic. In that case, the price of polluting inputs is sensitive to demand shocks. Expression (5.1) reveals also that a lower demand for polluting inputs in the home country raises the market price of polluting inputs only if $s_1^E > 0$. Indeed, a small economy, which features a share close to zero, will exert only a negligible impact on world-market prices.

The second term in the expression between square brackets on the RHS of (5.1) stands for the change in foreign demand for polluting inputs. In particular, environmental policies pursued by the home country may raise the demand for polluting inputs by country 3 through their negative impact on the market price of polluting inputs, P_E. These so-called leakage effects render domestic policies less effective in curbing global pollution.[51] The leakage effect can be derived from the model of country 3. In particular, by linearizing the factor-demand equation for polluting inputs and assuming a zero cross-price elasticity ϵ_{EY}, one derives for polluting input demand by country 3:

$$\tilde{E}_3^D = -\epsilon_{EE} \tilde{P}_E \qquad (5.2)$$

---

[51] Environmental policies in the home country may also cause relocation of energy-intensive industries. These leakage effects can be prevented by imposing proper border-tax adjustments (see Tang et al., 1998). The model in this chapter does not include leakage effects due to relocation of industries.

where $\epsilon_{EE} > 0$ stands for the direct price elasticity of polluting input demand in country 3. By substituting (5.2) into (5.1), one arrives at the following price equation for polluting inputs:

$$\tilde{P}_E = \phi \tilde{E}_1^D \tag{5.3}$$

$$where \quad \phi = \frac{\gamma s_1^E}{1 + \epsilon_{EE} \gamma (1 - s_1^E)} \tag{5.4}$$

The parameter ϕ in (5.3) measures the overall impact of changes in the demand for polluting inputs in country 1 on the world-market price of polluting inputs. According to (5.4), this elasticity is determined by three elements. First, it depends on the inverse-supply elasticity of polluting inputs, γ, which measures how the price of polluting inputs responds to changes in global demand for polluting inputs. Second, ϕ is related to the share of polluting inputs in the home country in terms of global demand for polluting inputs, s_1^E. This share measures the degree of monopsony power of the home economy on the market for polluting inputs. Finally, the elasticity in (5.4) depends on the demand elasticity of polluting inputs in country 3, i.e. the leakage effect. In particular, if leakage effects are important (ϵ_{EE} is large), domestic policies will have only a small effect on the price of polluting inputs.

The price of intermediate inputs
By linearizing the factor-demand equation for intermediate inputs in country 3, one obtains for the price of intermediate inputs (assuming $\epsilon_{EY} = 0$):

$$\tilde{P}_Y = -\frac{1}{\epsilon_{YY}} \tilde{Y} \tag{5.5}$$

where subscripts and superscripts for Y have been eliminated (since supply of intermediate inputs by country 1 equals demand by country 3). The demand elasticity for intermediate inputs in country 3, $\epsilon_{YY} > 0$, can be interpreted as the export elasticity for country 1. In particular, if the demand for intermediate inputs in country 3 is elastic, the export elasticity for the home country is large. In that case, the terms of trade will hardly respond to policies in country 1. If ϵ_{YY} is small, however, a reduction in exports from the home country exerts a large positive impact on the price. In that case, the home country possesses substantial monopoly power on the market for its export products.

Differences with the benchmark model

The linearized model of country 1 is largely equivalent to the linearized benchmark model from Table 3.1. There are two major differences, however. First, I add the price equations (5.3) and (5.5) to the model. Second, these endogenous prices of polluting inputs and intermediate inputs appear in the following three expressions:

Labor demand $$\tilde{L} = \tilde{E} - \sigma_{LE}[\tilde{W} + \tilde{T}_L - (1 - \theta_E)\tilde{P}_E - \tilde{T}_E] \qquad (5.6)$$

Non-profit condition $$\tilde{P}_Y = w_L(\tilde{W} + \tilde{T}_L) + w_E[(1 - \theta_E)\tilde{P}_E + \tilde{T}_E] \qquad (5.7)$$

Balance of payments $$\tilde{P}_Y + \tilde{Y} = w_C\tilde{C} + w_E(1 - \theta_E)(\tilde{P}_E + \tilde{E}) \qquad (5.8)$$

5.1.3   Welfare

In a similar way as in chapter 3, the welfare implications of an environmental tax reform are measured by the marginal excess burden. For each country, this marginal excess burden can be derived by linearizing the equations in Table 5.1 and adopting the same procedure as in appendix 3B to derive the compensating variation. Dividing the compensating variation by output in that country, one derives the marginal excess burden.

Welfare in the home country

For the home country, this procedure yields the following expression for the marginal excess burden:

$$MEB_1 = -(1-\theta_L)w_L\tilde{W}_1 - \frac{u_M}{\lambda}w_M\tilde{M}_1$$

$$\underbrace{\qquad\qquad}_{blue\ dividend} \qquad \underbrace{\qquad\qquad}_{green\ dividend}$$

(5.9)

where w_L and w_M denote shares for country 1. The marginal excess burden in (5.9) shows that welfare rises if private incomes in the home country increase (the blue dividend) and if environmental quality improves (the green dividend). By combining the government budget constraint (T3.8), the non-profit condition (5.7), the linearized expression for environmental quality, and expressions (5.2), (5.3) and (5.5), the MEB_1 can alternatively be written as:

$$MEB_1 = \underbrace{-\theta_L w_L \tilde{L}_1}_{(1)} - \underbrace{[\theta_E - \frac{u_M\gamma_{E1}}{\lambda}]w_E\tilde{E}_1^D}_{(2)} + \underbrace{[(1-\theta_E)w_E\tilde{P}_E - \tilde{P}_Y]}_{(3)} - \underbrace{\frac{u_M\gamma_{E3}}{\lambda}\epsilon_{EE}\tilde{P}_E}_{(4)}$$

(5.10)

The marginal excess burden in (5.10) contains four terms:

(1)   A labor-market distortion associated with distortionary labor taxes in country 1.

(2) An environmental distortion, measured by the difference between the initial pollution tax and the marginal environmental damage from domestic pollution.

(3) A terms-of-trade effect. A lower price of polluting inputs or a higher price of intermediate inputs reduces the price of imported goods (polluting inputs) relative to the price of exported goods (intermediate inputs). Hence, the terms of trade improves. This raises private incomes in the home country and, therefore, domestic welfare.

(4) Leakage effect. A lower market price of polluting inputs induces a higher demand for polluting inputs by country 3. If pollution crosses international borders, this deteriorates environmental quality in the home country, thereby

reducing green welfare. The leakage effect is measured by the elasticity $\gamma_{E3} \equiv m_{E3}E_3 / P_Y Y$, where $m_{E3} \equiv \partial M_1 / \partial E_3$.

The first two terms are equivalent to those in the benchmark model (see expression (3.10)). The last two terms originate in the endogenous prices of polluting inputs and intermediate inputs. Indeed, a change in the prices of polluting inputs and intermediate inputs causes a direct welfare effect for country 1 through the terms-of-trade effect (3). In this second-best framework, such changes in the terms of trade may affect welfare also indirectly through interactions with the labor-market distortion (1) or the environmental distortion (2). Changes in the price of polluting inputs affect welfare in country 1 also through the leakage effect (4).

Welfare in other countries
Following the same procedure as for the home country, one finds the following expressions for the marginal excess burdens of, respectively, country 2 and country 3:

$$MEB_2 = -\tilde{P}_E \tag{5.11}$$

$$MEB_3 = \beta_E \tilde{P}_E + \beta_Y \tilde{P}_Y \tag{5.12}$$

here β_E and β_Y stand for the cost shares of polluting inputs and intermediate inputs in value added in country 3. Expressions (5.11) and (5.12) contain two terms-of-trade effects. First, a lower price of polluting inputs improves the terms of trade in country 3, but is associated with a terms-of-trade loss in country 2. Hence, a lower price of polluting inputs redistributes welfare from country 2 towards the other two countries. Second, a higher price of intermediate inputs causes a terms-of-trade loss in country 3 and thus redistributes welfare from country 3 towards the home country. Since countries 2 and 3 feature no preexisting distortions in their economies, the marginal excess burdens in these countries is determined by the terms-of-trade effects alone.

Global welfare

I define global welfare (*MEB*) as the weighted sum of the marginal excess burdens of the three countries, where weights are determined by the shares of value added in each country in terms of global value added. This yields the following expression for global welfare:[52]

$$MEB = -\eta_1 [\theta_L w_L \tilde{L}_1 + [\theta_E - \frac{u_M \gamma_{EI}}{\lambda}] w_E \tilde{E}_1^D + \frac{u_M \gamma_{E3}}{\lambda} \epsilon_{EE} \tilde{P}_E] \qquad (5.13)$$

where η_1 denotes the share of value added of country 1 in terms of global value added. According to expression (5.13), changes in the world-market prices of polluting inputs or intermediate inputs do not directly affect global welfare through their impact on the terms of trade. Indeed, through the addition of the marginal excess burdens of the three countries, the terms-of-trade effects cancel out. Accordingly, changes in the terms of trade will affect the distribution of global welfare across different regions, but will leave the overall level of welfare unchanged.

If the economy in the home country would not be distorted initially (i.e. $\theta_L = \theta_E = u_M = 0$), then the marginal excess burden for the global economy in (5.13) would be zero. Hence, the introduction of a pollution tax in the home country would leave global welfare unchanged. Indeed, the welfare gain of the terms-of-trade improvement in country 1 would be exactly offset by a decline in welfare in other countries that experience a terms-of-trade loss. Accordingly, a zero pollution tax would be optimal from a global perspective.

In the presence of preexisting distortions in country 1 (without distortions in the other counties), expression (5.13) reveals that an environmental tax reform may exert an impact on global welfare through general equilibrium interactions with these prior distortions. In particular, (5.10) shows that welfare in country 1 rises with employment (due to the existing labor-market distortion), a higher price of polluting inputs (due to smaller leakage effects) and changes in the

---

[52] This expression ignores distributional issues. Effects on the distribution of welfare can be neutralized by appropriate lump-sum transfers across countries. The sign of the expression in (5.13) indicates whether a policy is potentially Pareto improving.

demand for polluting inputs in the home country (due to the environmental distortion). If changes in the terms of trade alleviate these distortions in country 1, then global welfare rises because the positive welfare effect in country 1 exceeds the negative welfare effect in the other countries.[53]

5.2   Tax shifting to foreign suppliers of polluting inputs

This section explores the welfare implications of an environmental tax reform in country 1 if the price of polluting inputs is affected by such a policy, i.e. $\phi > 0$. In order to focus on changes in the price of polluting inputs, this section assumes that the price of intermediate inputs, P_Y, is fixed because $\epsilon_{YY} \rightarrow \infty$. Subsection 5.2.1 discusses the optimal environmental tax from the point of view of the home country. Subsection 5.2.2 explores the welfare implications for the home country of a unilateral environmental tax reform, starting from an equilibrium that is not necessarily optimal. Finally, subsection 5.2.3 elaborates on the welfare implications for other countries and derives the optimal tax from a global perspective.

5.2.1   Optimal environmental tax

The model is solved along the lines of appendix 3C, where I used (5.3) and replaced (T3.2), (T3.3) and (T3.11) by, respectively (5.6), (5.7) and (5.8). Table 5.2 presents the reduced-form equations for employment, the demand for polluting inputs and private income in the home country. Note that, if $\phi = 0$, the solutions in Table 5.2 are equivalent to those from the benchmark model in Table 3.2.

The optimal tax on polluting inputs can be derived by first substituting the reduced forms for employment and polluting inputs from Table 5.2 into the expression for the MEB_1 in (5.10). Subsequently, this reduced form for the MEB_1

---

[53] If other countries would also feature preexisting distortions, and terms-of-trade losses would exacerbate these distortions, this result would not necessarily hold.

should be set equal to zero in order to derive the welfare optimum. Following this procedure, one derives the following expression for the optimal environmental tax:

$$\theta_E = \frac{1}{1 + \phi} [\phi + \frac{1}{\eta} \frac{u_M}{\lambda} (\gamma_{EI} - \frac{\gamma_{E3} \epsilon_{EE}}{w_E} \phi)] \tag{5.14}$$

$$\eta = \frac{1}{1 - T_L \eta_{LL}} \tag{5.15}$$

where η in (5.15) denotes the marginal cost of public funds (MCPF). The optimal pollution tax in the benchmark model is equal to the marginal environmental damage from domestic pollution, divided by the marginal cost of public funds (see (3.18)). In particular, with a MCPF larger than unity, the government less than fully internalizes the environmental externality, since preexisting tax distortions raise the social costs of environmental protection (see chapter 3). Expression (5.14) modifies this result in two directions. First, it reveals that the marginal environmental damage from domestic pollution will be smaller due to leakage effects (see the last term on the RHS of (5.14)). In particular, a lower level of pollution in the home country may induce higher pollution elsewhere, so that the marginal welfare gain associated with less local pollution becomes smaller. Accordingly, leakage effects call for a lower pollution tax in the home country.[54]

A second difference with the benchmark model is that the optimal environmental tax in (5.14) contains a term that is independent of the marginal

---

[54] If $\gamma_{EI} w_E < \gamma_{E3} \epsilon_{EE} \phi$, the deterioration of environmental quality on account of the leakage effect dominates the improvement in environmental quality due to less pollution in the home country. This may occur if the production process in country 3 causes more environmental damage per unit of polluting input than it does in country 1. This situation can be realistic for carbon taxes pursued by developed countries, since production methods in less developed countries are often more polluting. The rest of this section assumes that pollution taxes improve environmental quality, however.

environmental damage from pollution (see the first term between square brackets on the RHS of (5.14)). Hence, even without environmental considerations, i.e. $u_M = 0$, it is optimal for the government in country 1 to levy taxes on polluting inputs, i.e. $\theta_E = \phi/(1+\phi) > 0$. The reason is that, by taxing imported polluting inputs, the government in the home country changes the terms of trade in its favor (if $\phi > 0$). This terms-of-trade improvement boosts private incomes and thus domestic welfare. Intuitively, whereas individual firms are unable to behave strategically in demanding polluting inputs, country 1 as a whole might do so by reducing the domestic demand for polluting inputs, thereby lowering the price of polluting inputs. Hence, the terms-of-trade effect calls for a positive pollution tax. This finding reflects the well-known result from international trade theory that a country that possesses monopsony power on a world input market raises its own welfare by imposing an import tariff. On the one hand, such a tariff harms welfare in the home country by distorting the input mix. On the other hand, a tariff enhances welfare through a terms-of-trade improvement. The optimal tariff exactly equalizes the national costs and benefits. In my model, the pollution tax operates as an implicit import tariff. Indeed, I define $\theta_E = \phi/(1+\phi)$ as the optimal import tariff on polluting inputs.

Expression (5.14) reveals that the optimal tariff, represented by the first term on the RHS of (5.14), does not depend on the pre-existing labor-market distortions in the home country (i.e. on the MCPF). Hence, the second-best optimal tariff is equal to its first-best level. The reason is that pre-existing labor-market distortions magnify both the social costs of pollution taxes (in terms of a distorted input mix) and the national benefits of these taxes (in terms of an improvement in the terms of trade). Starting from an optimum, these costs and benefits exactly coincide so that the optimal level of the tariff remains unchanged.[55]

---

[55] This result is consistent with the findings by Boadway et al. (1973) and Williams (1998a). This latter study investigates interactions between trade- and labor-market distortions and finds that the optimal second-best tariff on an imported consumption commodity is equal to its first-best level if two conditions are met. First, the imported commodity should be an equally good substitute for leisure as the other commodities are. Second, there should be no rents from a fixed factor in production. In this chapter, these conditions are met.

This contrasts with the Pigovian component in (5.14), which falls with the MCPF. Intuitively, the social costs of internalizing externalities are borne privately and thus reduce labor supply through a lower consumer wage. In contrast, the social benefits of a cleaner environment are public, and leave labor supply unchanged. Pre-existing labor taxes thus magnify the social costs of internalizing externalities, but do not affect the social benefits. Accordingly, large distortions on the labor market call, *ceteris paribus*, for a lower optimal pollution tax (see also chapter 3).

The role of pollution taxes as a trade instrument may render the optimal environmental tax larger than the Pigovian rate (where $\theta_E = u_M \gamma_{EI}/\lambda$). However, a high MCPF and leakage effects reduce the optimal pollution tax. Hence, the optimal pollution tax is larger than the Pigovian rate only if terms-of-trade effects are relatively important, while leakage effects and preexisting distortions on the labor market are relatively small.

5.2.2   Environmental tax reform

Starting from an optimal environmental tax, changes in this tax do not improve welfare. Once one starts from a sub-optimal tax system, however, a marginal change in the tax structure does affect overall welfare. This section explores under which conditions welfare improves on account of both a blue dividend and a green dividend.

Starting from an equilibrium without environmental taxes (i.e. $\theta_E = 0$), the first row of Table 5.2 reveals that a shift away from labor taxes towards environmental taxes raises wages and employment if $\phi > 0$. This contrasts with the findings from chapter 3, where introducing environmental taxes leaves employment and private incomes unchanged. The underlying reason for this result is that the reduction in the demand for polluting inputs reduces the market price of polluting inputs (see expression (5.3)). This partly offsets the increase in the producer price induced by the pollution tax. The benefits of the lower price of polluting inputs accrue to workers in terms of a smaller decline in labor productivity. Together with the lower labor tax, the higher pollution tax thus results in a higher after-tax wage rate. Intuitively, part of the incidence of the environmental tax is shifted onto foreign suppliers of polluting inputs. Indeed, as the price of imported polluting inputs falls, country 2 shares in the economic burden of the environmental tax imposed by country 1. For the domestic economy, this means that the overall tax burden on labor declines.

Table 5.2: Reduced forms in the model with an endogenous price of polluting inputs (variables for the home country; subscripts and superscripts eliminated)

	$\tilde{T}_E$
$\Delta \tilde{L}$	$-w_E \dfrac{[\theta_E - (1-\theta_E)\phi]}{[1 + (1-\theta_E)\phi \frac{\sigma_{LE}}{w_L}]} \dfrac{\sigma_{LE}}{w_L} \eta_{LL}$
$\Delta \tilde{E}$	$-\dfrac{(1-\theta_L)w_L - \eta_{LL}\theta_L w_L}{1 + (1-\theta_E)\phi \frac{\sigma_{LE}}{w_L}} \dfrac{\sigma_{LE}}{w_L}$
$\Delta \tilde{w}$	$-w_E \dfrac{[\theta_E - (1-\theta_E)\phi]}{[1 + (1-\theta_E)\phi \frac{\sigma_{LE}}{w_L}]} \dfrac{\sigma_{LE}}{w_L}$

where $\Delta = (1 - \theta_L)w_L - \eta_{LL}[\theta_L w_L + w_E \dfrac{[\theta_E - (1-\theta_E)\phi]}{[1+(1-\theta_E)\phi\frac{\sigma_{LE}}{w_L}]}] > 0$

If one imposes an environmental tax reform in an equilibrium with positive environmental taxes (i.e. $\theta_E > 0$), Table 5.2 reveals that the effects on private income and employment are ambiguous. This is because of two opposing forces. On the one hand, the reform causes a tax burden effect. Through this channel, employment and private income drop (see chapter 3 for an explanation). On the other hand, wages and employment rise through the channel of the terms-of-trade improvement, caused by the lower market price of polluting inputs. On balance, starting from positive environmental taxes, an environmental tax reform yields a blue dividend only if the terms-of-trade effect dominates the tax burden effect. This happens if $\theta_E < \phi/(1+\phi)$ i.e. if the initial environmental tax rate is below the optimal import tariff, defined in the text below (5.14). If this condition holds, the efficiency cost associated with a less efficient input mix is offset by the terms-of-trade improvement caused by a lower input price. Hence, further increases in the pollution tax improve blue welfare.[56] If the initial pollution tax exceeds the optimal import tariff, i.e. $\theta_E \geq \phi/(1+\phi)$, however, the blue dividend fails. Indeed, the government then faces a trade-off between the non-environmental costs of further increases in the pollution tax (measured by the difference between the efficiency cost of a more distorted input mix and the terms-of-trade improvement) and the marginal environmental benefits from less pollution in the home country (corrected for the leakage effect).

5.2.3   Optimal tax from a global perspective

A lower demand for polluting inputs in country 1 reduces the world-market price of polluting inputs if $\phi > 0$. This lower price improves the terms of trade in countries 1 and 3, but harms the terms of trade in country 2. Accordingly, a lower demand for polluting inputs in country 1 causes a redistribution of welfare from country 2 towards countries 1 and 3 (see (5.10), (5.11) and (5.12)). To illustrate, if European countries unilaterally impose a tax on fossil fuels, the energy-producing countries would suffer from lower incomes, but the US may experience a welfare gain since the lower energy price improves their terms of trade.

---

[56] In the special case that the supply of polluting inputs is completely inelastic, i.e. ϕ is infinitely large, a higher pollution tax always raises employment and private income in the home country. However, the double dividend fails because pollution does not fall. Indeed, a large value for ϕ renders pollution taxes efficient instruments for trade policy, but ineffective instruments for environmental protection.

According to (5.13), the terms-of-trade effects have no direct implications for global welfare. However, in the second-best framework of this chapter, they may raise global welfare indirectly through interactions with preexisting distortions in the home country. This is illustrated by the optimal pollution tax from a global point of view. In particular, by substituting the reduced forms for employment and polluting inputs from Table 5.2 into the expression for the *MEB* in (5.13) and setting it equal to zero, one derives the following optimal environmental tax from a global perspective:

$$\theta_E = \frac{1}{1 + \phi T_L \eta_{LL}} [\phi T_L \eta_{LL} + \frac{1}{\eta} \frac{u_M}{\lambda} (\gamma_{EI} - \frac{\gamma_{E3} \epsilon_{EE}}{w_E} \phi)] \qquad (5.16)$$

The optimal pollution tax in (5.16) is a modification of the national optimal tax in (5.14). In particular, the first term between square brackets on the RHS of (5.16) measures the optimal tariff, i.e. the optimal tax from a non-environmental point of view. This tariff depends on the magnitude of preexisting tax distortions in the labor market in country 1 (i.e. $T_L \eta_{LL}$). If distortions are absent (i.e. $T_L = 0$), there is no reason for the global authority to levy a tariff on polluting inputs in country 1 for non-environmental reasons. Indeed, the welfare improvement in country 1 due to the terms-of-trade improvement exactly coincides with the welfare loss in other countries. However, if the labor market in country 1 is distorted initially (i.e. $T_L > 0$), the optimal tariff is positive from a non-environmental point of view. Indeed, the welfare gain in the home country exceeds the welfare loss in other countries. This is because small pollution taxes in the home country boost employment as positive terms-of-trade effects raise the consumer wage and, therefore, labor supply.[57]

The second term between square brackets on the RHS of (5.16) stands for the marginal environmental damages from pollution in country 1. In particular,

---

[57] This result holds only if the global authority has no access to more efficient instruments to redistribute welfare to country 1. Furthermore, an important assumption underlying this result is that initial distortions in the other two countries are absent.

in optimizing the pollution tax, the global authority incorporates the environmental externality in country 1.

A comparison of (5.16) with (5.14) shows that the national optimal tariff typically exceeds the global optimal tariff. The reason is that, in optimizing the import tariff, individual nations incorporate the welfare gain associated with income redistribution across countries through terms-of-trade effects. A global optimizer does not incorporate these redistributional effects.

5.3   Tax shifting to foreign users of intermediate inputs

This section investigates how policies pursued in the home country affect the price of intermediate inputs, P_Y. This section assumes that the price of polluting inputs is fixed, i.e. $\phi = 0$. Subsection 5.3.1 derives the optimal environmental tax for the home country. Subsection 5.3.2 discusses the consequences of an environmental tax reform if one starts from a sub-optimal tax system in the home country. Finally, subsection 5.3.3 derives the optimal tax from a global point of view.

5.3.1   Optimal environmental tax

The model with an endogenous price of intermediate inputs is solved along the lines of appendix 3C, where (T3.3) and (T3.11) are replaced by (5.7) and (5.8) and I used (5.5). Table 5.3 presents the reduced forms for employment, the demand for polluting inputs and private income. If $\epsilon_{YY} \to \infty$, the reduced forms in Table 5.3 are equivalent to those in Table 3.2.

The optimal tax on polluting inputs can be derived by substituting the reduced forms for labor, pollution and the output price into (5.10) and setting the MEB_1 equal to zero. Thus, one finds:

$$\theta_E = \frac{1}{\epsilon_{YY}} + (1 - \eta_{LL}T_L + \frac{(1 + T_L)\eta_{LL}}{\epsilon_{YY}}) \frac{u_M \gamma_E}{\lambda} \tag{5.17}$$

The optimal tax in (5.17) differs from that in (3.18) in two respects. First, the optimal pollution tax in (5.17) contains a term that is not related to the marginal environmental damage from pollution (see the first term on the RHS of (5.17)). Indeed, in the absence of environmental considerations, i.e. $u_M = 0$, the optimal pollution tax is positive if the export elasticity is less than infinity. The intuition behind this term is very similar to the optimal tariff in section 5.2.1. Indeed, whereas individual firms face a perfectly competitive output market, all domestic firms together have monopoly power on the international market for intermediate inputs. The government can exploit this power through raising the tax burden on the private sector. This reduces the production of intermediate inputs and increases the price of intermediate inputs, thereby improving the terms of trade of the home country. Accordingly, the tax burden is shifted onto foreigners. The pollution tax thus operates as an implicit tariff on exported commodities. The magnitude of this tariff depends on the export elasticity of country 1 alone: i.e., it is independent of preexisting tax distortions on the labor market (see section 5.2.1 for an explanation).[58]

A second difference between expressions (5.17) and (3.18) is that, if the export elasticity is less than infinity, environmental protection is less expensive. Intuitively, by distorting the input mix, pollution taxes typically reduce output. Since less output improves the terms of trade for the home country, this mitigates the costs of internalizing environmental externalities. Indeed, strong market power on export markets is associated with a smaller MCPF.

5.3.2   Environmental tax reform

This section turns to the welfare effects of an environmental tax reform if one starts from equilibria that are not necessarily optimal.

---

[58] It would be more efficient for the government to adopt an explicit tariff on intermediate inputs since this would not distort the input mix between labor and polluting inputs. This chapter assumes, however, that the government does not have access to such explicit tariffs, e.g. due to trade rules. In that case, expression (5.17) reveals that there may be a role for pollution taxes to operate as an implicit tariff on intermediate inputs.

If $\theta_E = 0$, the first row of Table 5.3 reveals that the introduction of an environmental tax boosts employment if $\epsilon_{YY} < \infty$. This is because the environmental tax lowers the demand for polluting inputs, reduces output and raises the price of exported intermediate inputs. This mitigates the decline in the value of the marginal product of labor. Accordingly, the before-tax wage drops less. Together with the lower tax on labor, this may yield a higher after-tax wage rate. The higher return to work stimulates labor supply. Intuitively, producers in country 3 share in the burden of the environmental tax imposed by the domestic country due to a change in the terms of trade. Consequently, the tax burden imposed on the private sector in the domestic economy declines. Hence, one obtains both a higher level of employment and higher private incomes if initial pollution taxes are zero.

If the initial environmental tax is positive (i.e. $\theta_E > 0$), Table 5.3 reveals that raising the environmental tax further can either raise or reduce employment and private income. This depends on the term $(\theta_E - 1/\epsilon_{YY})$. The tax term, θ_E, represents the tax burden effect, discussed in chapter 3. The second term, $1/\epsilon_{YY}$, measures the terms-of-trade effect that operates through the impact on the market price of intermediate inputs. Only if the terms-of-trade effect is sufficiently large compared to the tax burden effect does an environmental tax reform yield a blue dividend. From expression (5.17), one verifies that this happens if the pollution tax is smaller than the optimal tariff.

Table 5.3: Reduced forms in the model with an endogenous price of intermediate inputs (variables for the home country, subscripts and superscripts eliminated)

	$\tilde{T}_E$
$\Delta\tilde{L}$	$\dfrac{(\dfrac{1}{\epsilon_{YY}} - \theta_E)\dfrac{w_E}{w_L}\sigma_{LE}}{1 + \dfrac{1}{\epsilon_{YY}}\dfrac{w_E}{w_L}\sigma_{LE}}\eta_{LL}$

$$\Delta \tilde{E} \qquad -\frac{(1-\theta_L)w_L - \eta_{LL}(\theta_L - \dfrac{1}{\epsilon_{YY}})w_L}{1 + \dfrac{1}{\epsilon_{YY}}\dfrac{w_E}{w_L}\sigma_{LE}}\dfrac{\sigma_{LE}}{w_L}$$

$$\Delta \tilde{w} \qquad \frac{(\dfrac{1}{\epsilon_{YY}} - \theta_E)\dfrac{w_E}{w_L}\sigma_{LE}}{1 + \dfrac{1}{\epsilon_{YY}}\dfrac{w_E}{w_L}\sigma_{LE}}$$

$$\Delta = (1-\theta_L)w_L - \eta_{LL}[(\theta_L - \dfrac{1}{\epsilon_{YY}})w_L + (\theta_E - \dfrac{1}{\epsilon_{YY}})w_E \dfrac{1 - \dfrac{1}{\epsilon_{YY}}\sigma_{LE}}{1 + \dfrac{1}{\epsilon_{YY}}\sigma_{LE}\dfrac{w_E}{w_L}}]$$

5.3.3   Optimal tax from a global perspective

The terms-of-trade improvement in country 1 is associated with a loss in the terms of trade in country 3. Hence, the environmental tax reform involves a redistribution of welfare from country 3 towards the home country. Similar to section 5.2.3, this redistribution of welfare does not affect global welfare directly, but only through interactions with preexisting distortions in the home country (see (5.13)). By substituting the reduced forms for employment and pollution from Table 5.3 into (5.13), and setting the *MEB* equal to zero, one derives the following optimal pollution tax from a global point of view:

$$\theta_E = \frac{1}{1+(1+T_L)\eta_{LL}/\epsilon_{YY}}[\frac{T_L\eta_{LL}}{\epsilon_{YY}} + (1 - \eta_{LL}T_L + \frac{(1+T_L)\eta_{LL}}{\epsilon_{YY}})\frac{u_M\gamma_E}{\lambda}] \qquad (5.18)$$

The first term between square brackets on the RHS of (5.18) stands for the global optimal tariff, while the second term denotes the Pigovian component. Similar to the previous section, the optimal tariff modifies the optimal tariff from a national point of view in (5.17) by multiplying it with $T_L \eta_{LL}$. In particular, in contrast to national governments, the global optimizer does not incorporate the national welfare improvements in country 1 that are due to a redistribution of income across countries. Indeed, the global authority is interested only in the welfare improvements due to interactions with preexisting distortions in the home country. The previous section reveals that terms-of-trade improvements raise employment in the home country if $\theta_E < 1/\epsilon_{YY}$ and $\eta_{LL} > 0$. If $T_L > 0$, this expansion in employment raises welfare in country 1. In that case, the welfare gain in the home country exceeds the welfare loss in country 3. In the absence of other instruments to redistribute income between countries, the global optimizer will therefore set the pollution tax in the home country at a positive rate, but below the national optimal tax.[59]

Discussion
These findings are closely related to those of Buchanan (1969). For a closed economy, he finds that the optimal pollution tax is smaller than the Pigovian rate if the individual producer that causes the externality is a monopolist. Intuitively, by reducing output, the pollution tax exacerbates the preexisting distortion associated with monopoly power.

In my model, the home country possesses monopoly power on the international market for intermediate inputs. Hence, it is optimal for the nation as a whole to marginally reduce output in order to reap the monopoly rents in the form of terms-of-trade improvements. Indeed, the home country would operate as a monopolist if it imposes a tariff on polluting inputs equal to $\theta_E = 1/\epsilon_{YY}$. Starting from this initial situation and ignoring, for the moment, the distortionary tax on labor (i.e. $T_L = 0$), my model resembles that of Buchanan. In particular, the home country plays the role of the monopolist, while a global authority can

---

[59] This result may change if the terms-of-trade loss in other countries would exacerbate preexisting distortions in those countries.

be interpreted as the government that aims to internalize the environmental externality. In Buchanan's model, the monopoly already internalizes part of the environmental externality by producing an inefficiently low level of output. Indeed, the monopoly distortion operates as an implicit tax on pollution. Accordingly, the optimal pollution tax in Buchanan's model measures the *additional* tax on pollution that goes beyond the implicit pollution tax. This additional tax is typically smaller than the Pigovian rate. In my model, an explicit tax on pollution, i.e. the tariff, measures the preexisting world-wide distortion caused by the monopoly behavior of country 1. By reducing pollution, this tariff also internalizes part of the environmental externality. Accordingly, the additional tax on polluting inputs that should be levied by a global optimizer equals the difference between the marginal environmental damage from pollution and the initial pollution tax. Hence, the additional tax on polluting inputs is below the Pigovian rate.[60]

5.4   Numerical simulations

Table 5.4 illustrates how important terms-of-trade effects can be for the double dividend. In particular, it presents the effects of a 10% increase in energy taxes on firms in the home country, recycled through lower labor taxes. In the simulations, I either use the model of Table 5.1 with an endogenous price of energy, or the same model with an endogenous price of intermediate inputs. The model for the home country is calibrated on the basis of the same shares and parameter values as the benchmark model in section 3.5. The parameters that determine the terms-of-trade effects, ϕ and ϵ_{YY}, will be varied in the simulations. This also holds for the shares of other countries, since these may impact the terms-of-trade elasticities. Table 5.4 reports the effects of a reform on employment, pollution, the terms of trade and private income in the home

---

[60] If the monopoly distortion is large compared to the marginal environmental damage, the initial tariff on polluting inputs may exceed the Pigovian tax. In that case, global welfare would be optimized by subsidizing, rather than taxing, polluting inputs.

country, welfare in other countries, global welfare (from a non-environmental point of view) and global pollution.

5.4.1   Endogenous energy price

Expression (5.4) reveals that the value ϕ depends on three important variables. First, it is positively related to the share of energy consumption of the home country in terms of global energy consumption. In particular, if the home country represents a small economy, this share will be close to zero. To illustrate, the Netherlands consumes only 1% of global energy use (measured in millions of tons of oil equivalents). Hence, a unilateral energy tax in the Netherlands will hardly affect the terms of trade through changes in energy prices. European energy consumption amounts to approximately 16% of global energy consumption, while the share of Northern America is around 27%. The demand by the OECD countries together is almost 50% of global energy consumption (Ministry of Economic Affairs, 1989). Hence, a coordinated policy among EU countries or Northern America or all OECD countries together will potentially have important terms-of-trade effects.

The second variable determining the parameter ϕ is the price elasticity of energy supply, measured by the inverse of the parameter γ. In particular, the more elastic energy supply is (i.e. the smaller γ), the less energy prices will respond to demand shocks. Empirical estimates suggest that the supply elasticity may differ among various energy sources. To illustrate, Burnieaux et al. (1992b) find a price elasticity of oil supply between 1.0 and 3.0, and an elasticity of 5.0 for coal supply. Barns et al. (1992) find a supply elasticity of 1.0 for both coal and oil. Whalley and Wigle (1991), in their model, use a carbon-based energy-supply elasticity of 0.5. They also perform sensitivity analysis on this parameter, ranging from 0.1 to 1.5. In the simulations below, I take a price elasticity of the energy supply of 0.75, which implies that $\gamma = 1.33$.

The final variable that determines the terms-of-trade effect of a unilateral energy tax is related to the leakage effect. In the model of this chapter, this leakage depends on two variables. First, is the price elasticity of energy demand in countries that do not impose the energy tax. Chapter 4 cites a number of empirical studies from which I derive an energy-demand elasticity of 0.6. In the simulations discussed below, I take a value for the demand elasticity of foreign energy of -0.5. The second variable that determines the leakage effect is the size of the non-participating countries in terms of global energy consumption, i.e. $1 - s_1^E$.[61]

An OECD-wide energy tax reform

The empirical findings of the above-mentioned studies are used to derive a plausible value of the parameter ϕ in expression (5.4). In particular, first consider an OECD-wide energy tax and take $s_1^E = 0.5$. With $\gamma = 1.33$ and $\epsilon_{EE} = 0.5$, this implies a value for ϕ of 0.5. The effects of the OECD-wide energy tax reform are presented in the second column of Table 5.4. It reveals that energy demand within the OECD falls by 3.6%. This reduces the price of energy on the world market by almost 1.8%, thereby improving the terms of trade for the OECD countries. Accordingly, employment and private income increase by, respectively, 0.16% and 0.31%. Hence, the terms-of-trade effect dominates the tax burden effect so that the environmental tax reform yields a strong double dividend in the OECD area.

For the rest of the world (country 3), I assume that the production share of intermediate inputs is 0.25, while the energy share is 0.05. With these values, an environmental tax reform in the OECD reduces welfare in country 2 by 1.8%, while welfare in country 3 increases by 0.09%. On balance, the effect on global non-environmental welfare is slightly negative. Hence, although the pollution tax yields a double dividend for the OECD countries, it is not attractive from a global non-environmental perspective. However, the small non-environmental

---

[61] A number of empirical studies have investigated the importance of carbon leakage effects, thereby taking into account also the effect through relocation of energy-intensive industries. To illustrate, the OECD GREEN model suggests that carbon leakage effects are insignificant in case of an OECD-wide carbon tax: a CO2 reduction in the OECD area is accompanied by a rise in CO2 emissions in non-OECD countries by only 3% to 6% of that reduction in the medium term. In case of a unilateral energy tax in the EU, the model predicts a somewhat larger carbon leakage effect of 11% (see e.g. Burniaux et al., 1992a). Using CPB's WorldScan model, Tang et al. (1998) find a much larger carbon leakage effect for a European energy tax of almost 80%.

cost of the reform should be weighed against the reduction in global pollution. Indeed, I find that although energy consumption in country 3 increases by 0.45% due to the lower price of energy, global pollution falls by 1.34%. The carbon leakage effect in this simulation is approximately 25%.

An EU-wide energy tax reform
If the energy tax would be imposed only in the European Union, or possibly only in the Northern countries of the EU, the energy share of the home country in terms of global energy consumption would be much smaller. In particular, taking $s_1^E = 0.12$ and maintaining the elasticities of energy supply and energy demand, I arrive at $\phi = 0.1$. With the smaller value of ϕ, the third column of Table 5.4 reveals that employment and private incomes no longer rise, as was the case with an OECD-wide tax. Hence, an energy tax reform in the Northern EU countries is less likely to yield a double dividend than an OECD-wide energy tax. Furthermore, such a European energy tax reform involves less redistribution across regions. Indeed, welfare in oil-supplying countries falls by 0.47%, which is less than in the case of an OECD-wide tax. With a production share of energy of 0.05 in country 3, and a share of intermediate inputs of 0.03, welfare in country 3 rises by only 0.02%. Pollution in country 3 rises by 0.23% on account of the leakage effect, which is approximately 36%. On balance, global non-environmental welfare falls slightly, while global pollution drops by 0.36%.[62]

---

[62] Tang et al. (1998) find that a 25% energy tax in the EU reduces global emissions by 0.8%. This is slightly less than my simulations suggest, possibly because this chapter ignores relocation of industries. Tang et al. find that appropriate border-tax adjustment may reduce carbon leakage caused by relocation of industries, thereby implying reductions in global emissions that exceed those in my simulations.

Table 5.4: Effects of a 10% increase in pollution taxes on firms in country 1 if either the price of polluting inputs or the price of intermediate inputs is endogenous

	$\phi = 0$ $\epsilon_{YY} \to \infty$	$\epsilon_{YY} \to \infty$ $\phi=0.5$	$\epsilon_{YY} \to \infty$ $\phi=0.1$	$\phi = 0$ $\epsilon_{YY}=2$	$\phi = 0$ $\epsilon_{YY}=\frac{1}{2}$
Effects for the home country					
Employment	−0.07	0.16	0.00	0.18	−0.01
Pollution	−5.07	−3.57	−4.66	−4.65	−5.00
Terms of trade		1.79	0.47	0.37	0.08
Welfare[a]	−0.14	0.31	−0.01	0.39	−0.02
Effects for other countries					
Welfare in country 2		−1.79	−0.47	0	0
Welfare in country 3		0.09	0.02	−0.10	−0.02
Global welfare[a]		−0.01	−0.01	−0.01	−0.02
Global pollution		−1.34	−0.36	−2.32	−2.50

[a] Welfare viewed from a non-environmental point of view

5.4.2   Endogenous price of intermediate inputs

If $\epsilon_{YY} = 2$, a 1% reduction in domestic output, reduces the price of intermediate inputs by 0.5%. In that case, the fourth column of Table 5.4 reveals that the terms of trade improves on account of an environmental tax reform because energy taxes in country 1 indeed reduce the production of intermediate inputs. Accordingly, private income increases by 0.39%, while employment expands by 0.18%. At the same time, the reduction in domestic pollution is substantial (−4.65%). Hence, the domestic country reaps a double dividend since it succeeds in shifting the tax burden onto foreigners.

Using a production share of intermediate inputs in country 3 of 0.25, I find that welfare in country 3 falls by 0.1%. Also global non-environmental welfare falls by 0.01%. However, since there is no carbon leakage effect in this experiment, the lower level of pollution by the home country substantially reduces global pollution. In particular, taking a share of 0.5 for energy consumption of the home country (a share that is consistent with an OECD-wide tax), I find that global pollution falls by 2.3%.

If the export elasticity is larger (in absolute value), terms-of-trade effects are less important and the double dividend may fail (see the fifth column in Table 5.4). In that case, the adverse welfare effect on account of the tax burden effect in the home country dominates the favorable welfare effect associated with the improvement in the terms of trade. In that case, the welfare loss in country 3 is also smaller. Global welfare falls by 0.02% while the reduction in global pollution is 2.5%.

Empirical evidence for the Netherlands suggests that the export elasticity on a macro level is around 2 to 3 (see Draper, 1995). Hence, a value for ϵ_{YY} of 2 seems realistic for the Dutch economy. There are two reasons why this value may be too small, however. First, as suggested by CPB (1992), energy-intensive industries usually operate on highly competitive international markets. This would imply that the export elasticity in case of unilateral energy taxes should be set at a somewhat larger value (in absolute terms). Second, my model assumes that all domestically produced goods are exported to country 3. If the export share of domestic production would be smaller, the terms-of-trade effect would also be smaller. Hence, the simulation results in the fifth column might be more relevant for the Dutch economy than the results in the fourth column. If the home country would represent a larger economy, such as the EU or the OECD, the export elasticity is probably smaller than that for the Netherlands. In that case, the fourth column may be more realistic.

5.5   Conclusions

This chapter shows that an environmental tax reform may yield a double dividend if pollution taxes improve the terms of trade. In particular, if a country exerts monopsony power on the international market for polluting inputs, an import tariff is optimal because it improves the terms of trade of the home country, thereby raising private welfare. Environmental taxes on imported polluting inputs may operate as such a tariff. Similarly, a country that possesses monopoly power on output markets may employ environmental taxes to improve the terms of trade and thus shift the incidence of taxes towards foreigners that demand those products. In both cases, therefore, foreigners pay the costs of environmental protection in the home country.

A terms-of-trade gain for the home country is always accompanied by a terms-of-trade loss in another country. Hence, changes in the terms of trade typically redistribute welfare across countries but do not raise global welfare. In a second-best world, however, this chapter finds that terms-of-trade improvements in the home country not only raise welfare directly, but also indirectly through interactions with preexisting labor-market distortions. Accordingly, the welfare gain associated with improvements in the terms of trade in the home country typically exceed the welfare loss in other countries. Accordingly, a positive tariff can be optimal also from a global perspective. This result is conditional on the assumption that there are no other, more direct instruments to redistribute welfare across regions, e.g. due to international trade agreements. Furthermore, the model assumes that other regions in the world do not feature distortions in their economies.

Simulations with the model suggest that terms-of-trade effects are potentially important mechanisms in case of coordinated CO_2 policies. Indeed, for plausible parameter values, I find that an OECD-wide energy tax reform may yield a double dividend for the OECD due to favorable terms-of-trade effects. If the reform is implemented on a smaller scale, such as in the EU, the potential for a double dividend in the EU declines.

6   Environmental taxes and distributional concerns

This chapter explores the relationship between the double dividend and distributional concerns. The income distribution is generally important in the political debates on environmental tax reforms. Indeed, the welfare states in most Western European countries indicate that societies feature strong preferences for an equitable distribution of welfare among households. The benchmark model in chapter 3 is inappropriate to explore these distributional issues, since it starts from identical households.

To explore the double dividend and distributional concerns, this chapter modifies the household model from chapter 3 by introducing non-identical agents. In particular, households differ with respect to their skill level, which also causes differences in the labor income they receive. Society may feature preferences for an equitable distribution of welfare among the different households. The government may achieve this social objective through the provision of uniform lump-sum transfers or an allowance for tax payments on polluting commodities. Within this framework, this chapter illustrates the trade-offs between equity, efficiency and environmental externalities. In particular, an environmental tax reform may produce a double dividend by improving the quality of the environment and reducing the efficiency costs of taxation. However, such a double dividend is feasible only if the tax burden is shifted from high-skilled households (who receive relatively high labor income) towards low-skilled households (who rely more heavily on transfer income). This change in the distribution of welfare may reduce social welfare if the government features strong preferences for equity. Hence, despite a double dividend, social welfare may nevertheless fall. Alternative options for recycling the revenues from the environmental tax may yield more favorable outcomes from an equity point of view, but these reforms fail to yield a double dividend.

This chapter is related to other studies in the double-dividend literature. Bovenberg and de Mooij (1992) and Bovenberg and van der Ploeg (1994b) introduce household income from lump-sum transfers provided by the government. However, these studies maintain the assumption of a representative household so that distributional concerns are unimportant. Proost and van Regemorter (1995) adopt an applied general equilibrium model for Belgium with non-identical households to explore the double dividend issue. They define a social welfare function that contains a parameter for the degree of income inequality aversion. The authors find that, in the presence of strong social preferences for an equitable income distribution, recycling the revenues from pollution taxes in a lump-sum fashion may be more attractive than using the revenues to cut distortionary taxes on labor -- even though this latter strategy entails bigger efficiency gains. Bovenberg and van Hagen (1997) develop a model with low-skilled and high-skilled households and find that the potential redistributive effect of an environmental tax reform depends on both the effects on the income distribution, and on the different valuations of high-skilled and low-skilled households for an improved quality of the environment. In particular, if low-skilled households benefit more from a cleaner environment than do high-skilled households, this may allow the government to reduce the amount of income redistribution from high-skilled to low-skilled households, without affecting the distribution of welfare in a broader sense.

The tax-reform analysis performed in this chapter is linked also to the literature on optimal second-best pollution taxes in the presence of heterogenous households. This was first analyzed by Sandmo (1975) in the context of a commodity-tax system. He finds that the externality does not affect the optimal tax on clean goods, but enters only the optimal-tax formula for the polluting commodity in an additive manner. Pirttilä and Schöb (1996) also explicitly derive the optimal commodity tax structure in the presence of externalities and non-identical households. They emphasize the importance of the utility structure of individual households for the optimal tax results. In particular, distributional concerns do not affect the optimal tax structure derived by Sandmo if private consumption is weakly separable from leisure and if Engel curves are linear. However, if Engel curves are non-linear (so that poor households exert a relatively large demand for polluting commodities), distributional concerns call for a relatively low tax on polluting commodities. This distributional concern conflicts with the objective to internalize environmental externalities. Accordingly, there may be a fundamental trade-off between equity and internalizing the externality. Johansson (1995) performs a similar optimal-tax analysis by focusing on the distributional effects related to different valuations of environmental damages among agents. Mayeres and Proost (1997) derive the optimal tax structure in the presence of a linear income tax. Their model is used to analyze different policies to combat congestion-type externalities.

Adopting the self-selection approach to optimal taxation, Pirttilä and Tuomala (1996) analyze the optimal tax structure in the presence of non-linear income taxes as a means of income redistribution between high- and low-ability households. To ensure that high-ability households will not mimic the behavior of low-ability households (e.g. by working less), their approach imposes an incentive-compatibility constraint, the so-called self-selection constraint. Pirttilä and Tuomala find that, if environmental quality and leisure are complements in utility, the optimal pollution tax is lower than the marginal environmental damage. The reason is that a dirtier environment makes leisure less enjoyable for high-ability households, which discourages them to mimic the choice of low-ability households. Accordingly, lower pollution taxes relax the self-selection constraint of high-ability households. This raises social welfare, as it facilitates redistribution policy. Another result by Pirttilä and Tuomala is that they extend a well-known result from optimal tax theory by Atkinson and Stiglitz (1976). These latter authors have shown that, if leisure is separable from consumption in utility, the government should adopt only non-linear income taxes, and not commodity taxes, to raise revenue and redistribute income. Pirttilä and Tuomala extend this result by showing that, in the presence of externalities that are separable from private goods, the optimal tax on the externality-generating commodity should be modified with Sandmo's additive term to correct for the externality. Cremer et al. (1997) find a similar result by using a model that allows for more than two households. The analysis in this chapter is restricted to linear income taxes.

The rest of this chapter is organized as follows. Section 6.1 presents the modified household model as compared to the benchmark model. Furthermore, it discusses alternative measures to explore the welfare implications of public policies in the presence of heterogenous households. In section 6.2, we analyze the effects of environmental taxes if the government uses the revenues in two alternative ways. In the first experiment, all revenues are used to cut the distortionary labor tax. In the second experiment, the revenues are recycled in such a way that real non-labor incomes are indexed to the real after-tax wage rate. Section 6.3 presents some numerical simulations to illustrate the importance of distributional issues in the double-dividend debate. Finally, section 6.4 concludes.

6.1   The model with heterogeneous households

Compared to the benchmark model in chapter 3, this section presents a modified household model with non-identical households. Accordingly, we are able to analyze the potential for a double dividend if an environmental tax reform has implications for the distribution of welfare among different households. In order to focus on distributional issues, this chapter abstracts from inputs that pollute the environment: i.e. E is non-polluting and untaxed. Accordingly, environmental quality in this chapter is determined by the demand for polluting consumption commodities alone. Subsection 6.1.1 presents the structure of the modified household model. Subsection 6.1.2 discusses the linearized version of the model. Finally, section 6.1.3 presents a measure to explore the welfare effects of public policies.

6.1.1   Structure of the model

The modified household model

The economy is populated by N heterogenous households, indexed by $i = 1...N$, who differ with respect to their skill level, h^i. These skills are assumed to be fixed. For each $i = 1...N - 1$, we assume that $h^i > h^{i+1}$, so that the first household features the highest skill level, while h^N denotes the household with the lowest skill. The budget constraint for each individual household reads as follows:

$$Wh^i l_S^i + s = c^i + P_D d^i + T_D(d^i - d^*) \qquad i = 1,...,N \qquad (6.1)$$

where lower-case symbols denote individual variables. Labor supply by household i (l_S^i) is transferred into effective labor supply by multiplying it with the skill index (h^i). For each unit of effective labor supply, households receive a uniform after-tax wage (W). Because high-skilled households are more productive, they supply more effective labor units and thus receive a higher after-tax wage income than low-skilled households do. The price of clean consumption commodities is taken as the numeraire which is set equal to unity.

The government ensures a minimum standard of living for each household. Since the government cannot observe the skill level of individual households, e.g. due to informational asymmetries, it has no access to differentiated lump-sum transfers to meet its social objectives. Accordingly, it has to rely on uniform lump-sum transfers, s, to protect the incomes of those with low skills.[63] These transfers are not subject to the labor tax.

Each household i spends its entire income on clean (c^i) and polluting (d^i) consumption commodities. Polluting consumption commodities are taxed if they go beyond a particular threshold level, d^*, the so-called the tax-free pollution allowance (TFPA). This TFPA is uniform across households. If the demand for polluting consumption commodities by a particular household i is below the

---

[63] Hence, we do not allow for non-linear income taxation as, for instance, in Pirttilä and Tuomala (1996) and Cremer et al. (1997). Linear income taxes are typically easier to administer, as the government does not need to observe individual incomes.

TFPA, i.e. $0 < d^* < d^i$, this household does not pay pollution taxes but rather receives a subsidy for each polluting commodity below the threshold.[64]

To better understand the role of the TFPA, we can rewrite the household budget constraint in (6.1) as follows:

$$Wh^i l_s^i + s + T_D d^* = c^i + (P_D + T_D) d^i \qquad i = 1,...,N \qquad (6.2)$$

According to expression (6.2), the TFPA, d^*, multiplied by the pollution tax rate, T_D, can be interpreted as a virtual source of income that is uniform across households, i.e. independent of ability. Indeed, the virtual income from the TFPA has the same properties as the explicit transfer, s, and can be used as an instrument to achieve social objectives related to income distribution. If we interpret the TFPA as a source of income, the RHS of (6.2) reveals that all polluting consumption commodities are taxed by T_D. In the rest of this chapter, we adopt this alternative interpretation of the TFPA. In particular, we will classify the virtual income from the TFPA ($T_D d^*$) as part on non-labor income.

Each household features the same preference structure, denoted by $u^i = u[q(c^i,d^i), v^i, M, G]$, where leisure, v^i, is the complement of labor supply and M and G stand for, respectively, environmental quality and public consumption. We impose some restrictions on the utility function that are important for our results.[65] First of all, we assume that public consumption and environmental quality are weakly separable from private commodities and leisure. This condition ensures that changes in the supply of these public goods do not directly affect private behavior. Second, we assume that private consumption commodities are weakly separable from leisure so that changes in labor supply

---

[64] This tax-free pollution allowance was advocated by the WRR (1992) and included in the Dutch energy tax reform of 1996 for electricity and natural gas used for heating. In contrast to the TFPA analyzed in this chapter, the Dutch TFPA is not refundable for households with very low energy consumption. This implies that individual consumption levels of electricity and natural gas should be observable for the government, which is actually the case.

[65] These restrictions have been widely discussed in recent contributions on optimal pollution taxes in the presence of heterogenous households; see Johansson (1995), Pirttilä and Schöb (1996), Pirttilä and Tuomala (1996) and Cremer et al. (1997).

do not affect the allocation of consumption over clean and polluting commodities. Finally, the subutility function $q(.)$ is homothetic in its two arguments. This implies that each household spends an equivalent proportion of income on clean and dirty goods; i.e., the Engel curves are linear. From utility maximization, we derive the first-order conditions described by (3.5) - (3.7).

Aggregation

By summing up over all households, we derive the following economy-wide household budget constraint from (6.2):

$$WL_S + S + T_D D^* = C + (P_D + T_D)D \qquad (6.3)$$

where capitals stand for macro aggregates, which are defined as $L_S = \sum h^i l_S^i$, $C = \sum c^i$, $D = \sum d^i$, $S = Ns$ and $D^* = Nd^*$. Aggregate labor supply in (6.3) is thus expressed in effective labor units.

In the presence of transfers and a TFPA, the government budget constraint looks as follows:

$$P_G G + S = T_L WL + T_D(D - D^*) \qquad (6.4)$$

6.1.2   Linearization

The modified household model of subsection 6.1.1 is linearized along the lines of chapter 3 and presented in Table 6.1. The individual household budget constraint in (T6.1) reveals that expenditures on clean and dirty consumption are financed by three sources of income: labor income, transfer income and the virtual income from the TFPA. The relative change in wages, transfers and the value of the TFPA is the same for all households. However, the income shares, denoted by, respectively, ϕ_L^i, ϕ_S^i and ϕ_D^{*i}, differ among households. This is because high-skilled households receive higher wages.

Expression (T6.2) shows how individual labor supply responds to changes in real after-tax wages, real transfers and the real value of the TFPA. The uncompensated wage elasticity (η_{LL}^i) and the income elasticity (η_L^i) of labor supply, both defined at the end of Table 6.1, may differ among households. In accordance with empirical evidence, we assume that the elasticities η_{LL}^i and η_L^i are positive for all $i = 1...N$.

Expressions (T6.3) - (T6.8) define the linearized equations for aggregate variables from the household model. In particular, expression (T6.7) reveals that the real value of the TFPA depends on the level of the TFPA itself and the pollution tax. On the one hand, a higher pollution tax reduces the real value of any type of income, including the virtual income from the TFPA. On the other hand, pollution taxes raise the real value of the TFPA because the right to freely pollute becomes more valuable at higher tax rates. On balance, the second effect dominates, so that a higher pollution tax unambiguously raises the real value of the TFPA.

Table 6.1: Linearized household model with heterogeneous agents

---

Expressions per household ($i = 1... N$)

Household budget $\qquad \phi_L^i(\tilde{l}_S^i + \tilde{W}_R) + \phi_S^i \tilde{S}_R + \phi_D^{*i} \tilde{D}_R^* = \phi_C \tilde{c}^i + \phi_D \tilde{d}^i$ (T6.1)

Effective labor supply $\qquad \tilde{l}_S^i = \eta_{LL}^i \tilde{W}_R - \eta_L^i \phi_S^i \tilde{S}_R - \eta_L^i \phi_D^{*i} \tilde{D}_R^*$ (T6.2)

Aggregate expressions

Household budget $\qquad \phi_L(\tilde{L}_S + \tilde{W}_R) + \phi_S \ddot{S}_R + \phi_D^* \tilde{D}_R^* = \phi_C \tilde{C} + \phi_D \tilde{D}$ (T6.3)

Effective labor supply $\qquad \tilde{L}_S = \eta_{LL} \tilde{W}_R - \eta_L \phi_S \tilde{S}_R - \eta_L \phi_D^* \tilde{D}_R^*$ (T6.4)

Real after-tax wage $\qquad \tilde{W}_R = \tilde{W} - \phi_D \tilde{T}_D$ (T6.5)

Real value of transfer income $\qquad \tilde{S}_R = \tilde{S} - \phi_D \tilde{T}_D$ (T6.6)

Real value of the TFPA $\qquad \tilde{D}_R = \theta_D(\tilde{D}^* - \phi_D \tilde{T}_D) + \tilde{T}_D$ (T6.7)

Government budget

$\qquad \tilde{S} + \phi_D^* \tilde{T}_D + \theta_D \phi_D^* \tilde{D}^* = (1 + T_L)\phi_L \tilde{T}_L + \phi_D \tilde{T}_D + T_L \phi_L(\tilde{W} + \tilde{L}) + \theta_D \phi_D \tilde{D}$ (T6.8)

Notation

$$\eta_L^i \;=\; \frac{v^i}{1-v^i+\phi_L^i v^i} \quad \text{income elasticity of labor supply of household i}$$

$$\eta_{LL}^i \;=\; [\sigma_V - \phi_L^i]\eta_L^i \quad \text{uncompensated labor supply elasticity of household i}$$

$$\eta_{LL} \;=\; \sum_{i=1}^{N} \frac{h^i l_s^i}{L_s}\eta_{LL}^i \quad \text{aggregate uncompensated labor supply elasticity}$$

$$\eta_L \;=\; \sum_{i=1}^{N} \frac{h^i l_s^i}{L_s}\frac{\phi_s^i}{\phi_s}\eta_L^i = \sum_{i=1}^{N} \frac{h^i l_s^i}{L_s}\frac{\phi_D^{*i}}{\phi_D^*}\eta_L^i \quad \text{aggregate income elasticity}$$

$$\eta_{LL}^* \;=\; \eta_{LL} + \phi_L\eta_L \quad \text{compensated labor-supply elasticity}$$

Shares

$$\phi_L^i = Wh^i l^i/Wh^i\, l^i+s+T_D d^* \qquad\qquad \phi_s^i = s/Wh^i\, l^i+s+T_D d^*$$
$$\phi_D^{*i} = (P_D+T_D)d^*/Wh^i\, l^i+s+T_D d^*$$
$$\phi_L = WL/WL+S+T_D D^* \qquad\qquad \phi_s = S/WL+S+T_D D^*$$
$$\phi_D^* = (P_D+T_D)D^*/WL+S+T_D D^*$$
$$\phi_D = (P_D+T_D)D/WL+S+T_D D^* \qquad\qquad \phi_C = C/WL+S+T_D D^*$$
$$where \quad \phi_L^i+ \phi_s^i + \phi_D^{*i} = \phi_L + \phi_s + \phi_D^* = \phi_C + \phi_D = 1$$

6.1.3   Welfare

This section first derives a welfare measure for individual households to analyze the implications of public policies for the distribution of welfare among individuals. Second, it aggregates the individual welfare effects into a utilitarian social welfare measure. Together with the distributional effects, this utilitarian welfare measure yields insight into the trade-offs between equity, efficiency and environmental quality.

Distribution of welfare
For each individual household, we can derive the compensating variation. This amounts to the transfer that needs to be provided to that individual household in

order to keep its utility at the level before the policy shock. The compensating variation is derived in the same way as in appendix 3B. Dividing the compensating variation by individual household income, we find for the marginal excess burden for household i:[66]

$$meb^i = -\tilde{b}^i - \frac{u_M^i \phi_M^i}{\lambda^i} \tilde{M} \tag{6.5}$$

$$\tilde{b}^i = \phi_L^i \tilde{W}_R + \phi_S^i \tilde{S}_R + \phi_D^{i*} \tilde{D}_R^* \tag{6.6}$$

where $\phi_M^i \equiv M/(Wh^i l^i + s + T_D d^*)$ and λ^i denotes the marginal utility of income for household i. The meb^i for household i in (6.5) is similar to the MEB for the representative household in the benchmark model (see (3.11)). In particular, it contains a private component, related to after-tax household income defined in (6.6), and an environmental component. According to (6.6), the private component is determined by three terms: after-tax real wage income, real transfer income and the real virtual income of the TFPA. The welfare effect of a relative change in each of these incomes is measured by the respective income shares. Since these shares differ across households, a relative change in a particular type of income will have implications for the distribution of welfare over the different households. To illustrate, high-skilled people will benefit relatively more from a higher real wage rate than low-skilled people do because the high-skilled feature a larger labor-income share.

In an economy with non-identical households, the valuation of an improved environmental quality may also differ among individuals. Hence, changes in environmental quality have distributional consequences, perhaps because the externality affects particular income groups more than others, or because the environment is a luxury good that is valued more highly by high-income groups. In the rest of this chapter, we abstract from such distributional effects. In particular, we assume that the value households attach to a cleaner environment

---

[66] Note that in chapter 3, the MEB is defined as the compensating variation divided by aggregate output. In expression (6.5), the compensating variation is divided by individual household income.

is proportional to their marginal utility of income (i.e. u_M^i/λ^i is equivalent for different households i). Since λ^i is relatively small for high-skilled households, this condition requires that u_M^i is also smaller for the high skilled than for the low skilled.[67]

A utilitarian measure for social welfare
It is quite common in the literature on optimal taxation with heterogenous households to define a Bergson-Samuelson social welfare function that depends on the utilities of the different individuals, i.e. $f(u^1, u^2, ..., u^N)$, where u^i stands for the utility of household i. By linearizing this social welfare function and using the individual meb^i from (6.5) to measure the negative of the marginal change in utility of household i, we derive the following measure for the social marginal excess burden ($SMEB$):

$$SMEB = \alpha_1 meb^1 + \alpha_2 meb^2 + ... + \alpha_N meb^N \qquad (6.7)$$

where a positive $SMEB$ denotes a loss in welfare. The $SMEB$ in (6.7) is a weighted sum of the individual marginal excess burdens of all individuals. The expression allows for different social preferences concerning the distribution of welfare across non-identical households, determined by the weights, α_i. For instance, if the government features strong preferences for an equitable distribution of welfare, it assigns a relatively large weight to the meb^i of low-skilled households (i.e. α_N is relatively large) and a relatively small share to the meb^i of high-skilled households (i.e. α_1 is small).

This chapter assumes that the underlying social welfare function is a utilitarian function that simply adds the utilities of all households.[68] In that case,

---

[67] Although this condition seems rather strong, empirical information on the distributional effects of changes in environmental quality is scarce. In this chapter, the conditions imposed on u_M^i/λ^i are attractive since they allow for a simple rule to maintain the welfare distribution across households (see expression (6.9)). For an analysis of the distributional consequences of different valuations of environmental quality, see Bovenberg and van Hagen (1997).

[68] This social-welfare function implies that there is no underlying micro foundation for the presence of transfers or a TFPA. Indeed, this would require explicit preferences for an

the weights in the *SMEB*, α_i, are equal to the income shares of households in terms of aggregate household income, i.e. $\alpha_i \equiv (Whl^i + s + T_D d^*)/(WL + S + T_D D^*)$. By substituting (6.5) and (6.6) into (6.7) and using the income shares for α_i, we can write the *SMEB* as follows:

$$SMEB = -[\phi_L \tilde{W}_R + \phi_S \tilde{S}_R + \phi_D^* \tilde{D}_R^*] - \frac{N u_M}{\lambda} \phi_M \tilde{M} \qquad (6.8)$$

where u_M/λ denotes the value that each household assigns to environmental quality in terms of his income and $\phi_M \equiv M/(WL + S + T_D D^*)$. The *SMEB* in (6.8) can be interpreted as a utilitarian measure for welfare that features no distributional concerns. Indeed, the RHS of (6.8) implies only a trade-off between efficiency and the environment. In particular, the first term on the RHS of (6.8) between square brackets is a measure for efficiency determined by aggregate private income, i.e. the sum of real after-tax labor and non-labor incomes. We refer to an increase in aggregate private income as a *utilitarian blue dividend*. The second term on the RHS of (6.8) stands for the welfare improvement due to a change in environmental quality. We call this term the *utilitarian green dividend*. In the next sections, we will speak about a *utilitarian strong double dividend* if an environmental tax reform yields both a utilitarian green dividend and a utilitarian blue dividend.

Two options for revenue recycling
Utilitarian welfare in (6.8) involves a pure efficiency measure that does not take into account the welfare implications due to changes in the income distribution. By looking at the individual marginal excess burdens for the different households, however, we can investigate the implications for the distribution of welfare across individuals. Although we do not explicitly assign a value to distributional concerns in (6.8), combining this welfare measure with the

---

individual utilities of households.

distributional effects allows us to analyze the trade-off between equity and efficiency.[69]

The effects on the *SMEB* in (6.8) and the individual *meb^i* in (6.5) will depend on the way in which the revenues from the environmental tax are recycled, i.e. through lower labor taxes, higher transfers, or a higher level of the TFPA. We distinguish two cases concerning the revenue recycling.

Case 1

Non-labor incomes are fixed (in terms of clean commodities). Transfers (S) and the TFPA (D^*) are considered to be exogenous policy instruments. The revenues from the environmental tax will be recycled through a lower labor tax alone.

Case 2

Real non-labor incomes are indexed to real after-tax wages according to:

$$\tilde{S}_R = \tilde{D}_R^* = \tilde{W}_R \tag{6.9}$$

The indexation rule in (6.9) implies that the revenues from the environmental tax should be used for a combination of lower labor taxes, higher transfers and a higher TFPA. If non-labor incomes are indexed to real after-tax wages according to (6.9), we can rewrite private welfare in (6.6) as a function of the after-tax real wage rate alone:

$$\tilde{b}^i = \tilde{W}_R \tag{6.10}$$

---

[69] Lundin (1998) adopts the concept of Dalton-improving reforms to assess the welfare effects of an environmental tax reform in Sweden. This concept does not start from a social welfare function, but rather adopts a social ranking of households in terms of disposable income. In particular, a reform is said to be welfare improving if it redistributes the tax burden from poor to rich households, without changing the order of ranking. Lundin uses an extension of this concept in two directions. First, following Mayshar and Yitzhaki (1995), he takes also efficiency considerations into account. In particular, a redistribution of income from the rich to the poor is said to be welfare improving only if the marginal cost of public funds does not rise. Second, Lundin's concept involves a multidimensional ranking that takes into account not only disposable income, but also household size and access to public transportation.

Hence, according to (6.10), the effects on private income are identical for all households. Accordingly, we can explore whether an environmental tax reform allows for Pareto-improving reforms. Indeed, an environmental tax reform has implications for the income distribution only in case 1.[70]

6.2   Environmental tax reform

We solve the model under the assumptions outlined in cases 1 and 2 of the previous section. In case 1, the government has three exogenous policy instruments at its disposal: the pollution tax, the level of transfers and the TFPA. To meet its revenue requirements, the government endogenously adjusts the labor tax. The results are discussed in subsection 6.2.1. In the second case, non-labor incomes are indexed to the real after-tax wage rate according to (6.9). In that case, the pollution tax is the only exogenous policy instrument. The revenues of higher pollution taxes are recycled through both lower labor taxes and changes in transfers and the TFPA so that (6.9) holds. Subsection 6.2.2 discusses the results of this latter experiment.

6.2.1   Exogenous non-labor incomes

This section discusses the effects of changes in pollution taxes, income transfers and the TFPA on employment, polluting consumption and aggregate private income. In each of these experiments, the labor tax is used to meet the government budget constraint ex post. In Table 6.2, the first column presents the partial effect of higher pollution taxes, i.e. for a given level of real non-labor incomes. Pollution taxes, however, affect the equilibrium not only directly, but also indirectly through their impact on the real values of transfers and the income

---

[70] This result depends on the restrictions imposed on utility. In particular, the separability assumptions and the condition of homotheticity imply that consumption taxes are equivalent to uniform linear taxes on income. Accordingly, we can maintain the income distribution across N non-identical households by using only three instruments.

from the TFPA. The second and third columns report the effects of changes in the real values of these respective non-labor incomes.

Table 6.2: Reduced forms in the model with heterogenous households

	$\tilde{T}_D$	$\tilde{S}_R$	$\tilde{D}_R^*$
$\Delta\tilde{L}$	$-\eta_{LL}\,\theta_D\,\phi_C\,\phi_D\,\sigma_{CD}$	$-(1-\theta_D\phi_D)\phi_S\,\eta_{LL}^*$	$-(1-\theta_D\phi_D)\phi_D^*\,\eta_{LL}^*$
$\Delta\tilde{D}$	$-(1-T_L\eta_{LL})\phi_L\,\phi_C\,\sigma_{CD}$	$-(1+T_L)\phi_L\phi_S\,\eta_{LL}^*$	$-(1+T_L)\phi_L\phi_D^*\eta_{LL}^*$
$\Delta\tilde{B}$	$-\phi_L\,\theta_D\,\phi_C\,\phi_D\,\sigma_{CD}$	$-(T_L+\theta_D\phi_D)\phi_L\phi_S\eta_{LL}^*$	$-(T_L+\theta_D\phi_D)\phi_L\phi_D^*\,\eta_{LL}^*$

$$\Delta = [1 - \theta_D\,\phi_D]\,\phi_L - \eta_{LL}\,[T_L + \theta_D\phi_D]\,\phi_L$$

Higher non-labor incomes

The second and third columns of Table 6.2 reveal that higher real transfers or a higher real value of the TFPA exerts a negative impact on employment, pollution and aggregate private income. This is because the increase in non-labor incomes is financed by a higher distortionary tax on labor. This reduces the after-tax wage rate and discourages the incentives to supply labor. The adverse effect on labor supply is measured by the compensated wage elasticity of labor supply, i.e. $\eta_{LL}^* = \eta_{LL} + \eta_L\phi_L$. Indeed, whereas the lower after-tax wage rate induces a substitution effect from consumption towards leisure, the income effect is compensated for, due to an increase in non-labor income. The reduction in labor supply erodes the base of the labor tax so that higher marginal taxes on labor are required to meet the government budget constraint. These higher taxes reduce aggregate private income.

Pollution taxes affect the real value of transfer income and the virtual income from the TFPA. In particular, by raising the consumer price (see (T6.6)), pollution taxes reduce the real value of transfer income. In contrast, the real value of the TFPA is positively related to the pollution tax rate, since these taxes make the right to pollute freely more valuable (see (T6.7)). In exploring the implications of higher pollution taxes, we should include these indirect effects on the real values of transfer income and the virtual income from the TFPA.

Environmental tax reform

A reform from labor to pollution taxes affects the equilibrium through three channels. The first channel, represented by the first column of Table 6.2, involves the tax burden effect. In particular, higher environmental taxes typically reduce employment and aggregate private income if we start from an equilibrium with positive environmental taxes (i.e. $\theta_D > 0$). This is because, compared to labor taxes, pollution taxes are less efficient instruments from a non-environmental point of view. Indeed, by distorting the allocation of consumption between polluting and non-polluting commodities, environmental taxes erode the tax base more than labor taxes do, thereby raising the efficiency costs of taxation.

The second channel through which pollution taxes affect the equilibrium involves a reduction in the real value of transfer income (see (T6.6) in Table 6.1). According to the second column of Table 6.2, the lower real value of transfers boosts employment and private income. Intuitively, the incidence of the tax on polluting commodities is borne by both labor- and transfer incomes. In contrast, the labor tax bears on labor income alone. Accordingly, the swap of pollution taxes for labor taxes shifts the tax burden from labor income towards transfer income. This stimulates the incentives for labor supply and reduces the efficiency costs of taxation. We call this positive effect on employment and private income associated with a redistribution of the tax burden the *tax shifting effect*.

The third effect of environmental taxes on employment and private income denotes an opposite tax shifting effect, i.e. a shift in the tax burden from non-labor incomes towards labor incomes. In particular, expression (T6.7) in Table 6.1 reveals that pollution taxes always increase the real value of the TFPA.

According to the third column of Table 6.2, this higher real value of the TFPA reduces employment and aggregate private incomes.

The utilitarian strong double dividend

Overall, the introduction of pollution taxes may produce a utilitarian blue dividend under the following three conditions. First, the tax shifting effect through lower real income transfers should be sufficiently large. This requires a large aggregate income share of transfers in the initial equilibrium (i.e. ϕ_S is large). Second, the tax burden effect should be relatively unimportant. This holds if the initial pollution tax rate is small (i.e. θ_D is small) and substitution between clean and polluting commodities is difficult. Finally, the increase in non-labor income due to a higher real value of the TFPA should be relatively small. This requires a small TFPA in the initial equilibrium (i.e. ϕ_D^* is small). If these three conditions are met, an environmental tax reform raises aggregate private income. Since aggregate pollution is likely to fall, there is a potential for a utilitarian strong double dividend.

Distributional consequences

If the utilitarian strong double dividend holds, it is accompanied by a shift in the income distribution from low-skilled to high-skilled households. In particular, high-skilled households will experience a net gain from an environmental tax reform, as the benefits associated with lower labor taxes exceed the costs from higher pollution taxes. In contrast, low-skilled households, who rely more heavily on income transfers, will be worse off due to the environmental tax reform. Indeed, these households suffer more heavily from a reduction in real transfers than they gain from the higher after-tax real wage. Hence, the utilitarian strong double dividend comes at the cost of a less equitable income distribution.

This illustrates the trade-off between efficiency and equity. In particular, by shifting the tax burden away from labor income towards non-labor income, the government alleviates the inefficiencies associated with redistribution. However, one may wonder why the government has created these inefficiencies in the first place. Indeed, if the government adopts a utilitarian social welfare function, there is no reason to redistribute incomes from high-skilled households towards low-skilled households. The initial equilibrium is thus sub-optimal if transfer payments are positive. Therefore, by reducing the real value of non-labor incomes, an environmental tax reform moves the tax system closer to the non-environmental optimum. This move on the efficiency/equity trade-off can be obtained also by reducing transfers directly and using this money to cut the distortionary labor tax. Hence, the tax shifting effect is independent of environmental taxes.

The presence of transfers, however, should be explained by distributional concerns. Indeed, the utilitarian social welfare function does not provide a micro-economic explanation for the presence of transfers in the initial equilibrium. To incorporate such a micro foundation, we would need to include social preferences for equity in our analysis. In that case, tax shifting from labor to non-labor income will enhance efficiency in a narrow sense (i.e. without distributional concerns) but will typically harm social welfare through its adverse effect on the social value attached to equity. Hence, the welfare gain associated with the utilitarian strong double dividend should be weighed against the welfare loss due to a less equitable distribution of welfare among households. In general, it is not clear whether social welfare will improve or not. Indeed, if the initial equilibrium characterizes the optimal trade-off between equity and efficiency, marginal movements along the efficiency/equity trade-off have no effect on social welfare.

6.2.2   Indexation of non-labor incomes

This subsection discusses the effects of an environmental tax reform if non-labor incomes are indexed to the real after-tax wage rate according to (6.9). Hence, there will be no effects on welfare due to changes in the distribution of welfare among households. This experiment can thus be used to explore whether an environmental tax reform can be Pareto improving, i.e. make every household better off than in the initial equilibrium. The reduced-form coefficients of this experiment are presented in Table 6.3.

Table 6.3: Reduced forms in the model with heterogenous households and indexation of non-labor incomes to real after-tax wages

	$\tilde{T}_D$
$\Delta \tilde{L}$	$-[\eta_{LL} - (\phi_S + \phi_D^*)\eta_L]\theta_D \phi_C \phi_D \sigma_{CD}$
$\Delta \tilde{D}$	$-[1 - T_L \phi_L(\eta_{LL} - (\phi_S + \phi_D^*)\eta_L)]\phi_C \sigma_{CD}$
$\Delta \tilde{B}$	$-\theta_D \phi_C \phi_D \sigma_{CD}$

$$\Delta = [1 - \theta_D \phi_D] - [\eta_{LL} - (\phi_S + \phi_D^*)\eta_L][T_L + \theta_D \phi_D]\phi_L$$

Environmental tax reform

With indexation of real non-labor incomes to real after-tax wages, Table 6.3 reveals that the effect of an environmental tax reform on aggregate private income is similar to that in the benchmark model (compare the reduced forms in Table 6.2 and Table 3.2). Intuitively, the indexation rule implies the absence of the tax shifting effect, discussed in the previous section. Accordingly, pollution taxes affect the equilibrium only through the tax burden effect. This suggests that an environmental tax reform is unable to produce a utilitarian strong double dividend if non-labor incomes are indexed to wages. Indeed, if the government wants to maintain the income distribution, aggregate private income unambiguously falls.[71]

---

[71] In the benchmark model, the reduction in labor supply is measured by the uncompensated labor supply elasticity. In this chapter, the effect on labor supply is determined by the difference between the uncompensated labor-supply elasticity, η_{LL}, and the income elasticity of labor supply, η_L, premultiplied by $\phi_S + \phi_D^*$. This latter parameter is relevant, since not only the real after-tax wage rate but also non-labor incomes decline. The effect on labor supply in this chapter is not necessarily smaller, however, than in chapter 3. Indeed, the elasticities in Table

Discussion on the weak double dividend

Goulder (1995a) defines the weak double dividend as follows: "By using revenues from the environmental tax to finance reductions in marginal tax rates of an existing distortionary tax, one achieves cost savings relative to the case where the tax revenues are returned to taxpayers in a lump-sum fashion." Somewhat later he states that "the weak-double dividend is relatively uncontroversial." (p. 159). Our analysis confirms the presence of a weak double dividend if we adopt the utilitarian welfare function. Indeed, this is illustrated by the difference in outcomes of subsections 6.2.1 and 6.2.2. On the one hand, if we use all the revenues from the environmental tax to reduce labor taxes, aggregate private income may rise (see subsection 6.2.1). On the other hand, if we use part of the revenues to raise lump-sum transfers instead of reducing labor taxes, aggregate private income unambiguously falls (see subsection 6.2.2). However, the efficiency improvement in 6.2.1 comes at the expense of a less equitable distribution of income across households. If the government features strong preferences for equity, the weak double dividend can thus be associated with a decline in social welfare.

This result is consistent with Proost and van Regemorter (1995), who show that recycling the revenues from pollution taxes in a lump-sum fashion may be more attractive compared to reducing marginal tax rates if the government has strong aversion against income inequality. Bovenberg and van Hagen (1997) modify this result by allowing different households to assign different valuations for environmental damages. In particular, if poor households feature stronger preferences for environmental improvements, e.g. because local air pollution affects especially areas where poor people live, a redistribution of income from poor to rich households may be justified because the distribution of the environmental benefits may maintain the distribution of welfare in a broader sense, i.e. including the environmental benefits. Indeed, the distributional

---

6.3 are reduced-form elasticities. From the definitions at the end of Table 6.1, we derive η_{LL}^{i} − $(\phi_s^{i} + \phi_D^{*i})\eta_L^{i} = [\sigma_V - 1]\eta_L^{i}$. This term is equivalent to the definition of the uncompensated labor supply elasticity in the benchmark model. Hence, the reduced form for employment is negative if the substitution elasticity between leisure and consumption, σ_V, exceeds unity.

consequences of improvements in environmental quality may compensate for the change in the income distribution imposed by the environmental tax reform.

Revenue-raising vs. non-revenue-raising
If, along with the pollution tax, the government increases the TFPA, it protects the incomes of low- skilled households who face a large share of non-labor income. Indeed, increasing the TFPA has similar effects as increasing lump-sum transfers: it favors the low skilled relative to the high skilled. More specifically, if all revenues of an environmental tax are used to increase the TFPA, the reform boils down to an environmental tax, the revenues of which are recycled in a lump-sum fashion. This way of revenue-recycling is certainly not the most efficient way.

This result is consistent with Parry (1997) and Goulder et al. (1997). These authors find that revenue-raising environmental policy instruments involve smaller efficiency costs than instruments that do not raise revenue (such as quotas), because of the favorable consequences associated with the revenue-recycling effect. Indeed, non-revenue-raising instruments such as grandfathered pollution permits have similar effects on efficiency as taxes, the revenues of which are recycled in a lump-sum fashion. Our model finds that, if pollution taxes raise substantial revenue, there are potential efficiency improvements to be gained through a tax shifting effect from labor to non-labor incomes. In contrast, if pollution taxes do not raise revenue because the government raises also the TFPA, the tax shifting effect works in the opposite direction by shifting the tax burden from non-labor to labor incomes. This makes the tax system less efficient as an instrument to raise revenue. Accordingly, the efficiency costs of environmental policy are higher if the environmental policy instruments do not raise public revenue. Goulder et al. (1997) show that the difference between the efficiency costs can be substantial. In particular, for US policy to reduce sulfur dioxide, they find that the efficiency cost of non-revenue-raising instruments can be 38% higher than for revenue-raising instruments.[72]

---

[72] Fullerton and Metcalf (1996) argue that the key distinction is not between revenue-raising and non-revenue-raising instruments, but whether a particular policy creates entry barriers

6.3   Numerical simulations

This section illustrates the findings of this chapter by presenting some numerical simulations with the model. The shares and elasticities used for the calibration of the model are presented in Table 6.4. In particular, most aggregate shares and parameters are calibrated by using data from the benchmark model. Furthermore, we assume that the share of labor income in terms of aggregate household income is 0.7. The share of transfer income is 0.275, which is a rough approximation of the share of social benefits in terms of GDP in many EU countries. The share of virtual income from the TFPA is 0.025. This latter value is introduced for illustrative purposes only because, except for the Netherlands, most countries do not yet adopt a TFPA. The initial level of the TFPA is well below the average level of energy consumption so that energy is taxed on a net basis. The initial energy tax is 0.2 in terms of the consumer price of energy. The structural parameters from the benchmark model imply that the income elasticity of aggregate labor supply is equal to 0.1, while the uncompensated labor supply elasticity equals 0.23. For the compensated labor-supply elasticity, this implies a value of 0.3.

Table 6.5 presents the effects of a 10% increase in the consumer price of energy due to a higher energy tax. We discuss the effects on employment, pollution, real after-tax wages and aggregate private income under two alternative strategies with respect to revenue recycling. The first three columns of Table 6.5 reflect the three channels through which an environmental tax reform affects the equilibrium if transfers and the TFPA remain constant (see subsection 6.2.1). In particular, pollution taxes cause a tax burden effect (first column), a tax shifting effect away from income transfers (second column), and a tax shifting effect towards the virtual income from the TFPA (third column). The fourth column is the sum of the first three columns and gives the overall

---

through scarcity rents (see also chapter 9). In our model, the distinction between revenue-raising and non-revenue-raising instruments is important, since it affects the distribution between labor and non-labor incomes. In particular, non-labor incomes associated with the TFPA may be interpreted as scarcity rents.

effect of an environmental tax reform from labor towards energy if non-labor incomes are not indexed. The fifth column in Table 6.5 presents the effects of an environmental tax reform if non-labor incomes are indexed to real after-tax wages. The last two rows of Table 6.5 denote the effects on private income for two specific households, namely a high-skilled household and a low-skilled household. The high-skilled household features a labor-income share of 0.88, a share of transfer income of 0.11 and a share of virtual income from the TFPA of 0.01. For the low-skilled households, these shares are, respectively, 0.16, 0.77 and 0.07.

Table 6.4: Calibration of shares and parameters of the model with non-identical households

Shares	Taxes	Elasticities
$\phi_L = 0.700$	$T_L = 1.0$	$\sigma_V = 3.00$
$\phi_S = 0.275$	$\theta_D = 0.2$	$\sigma_{CD} = 0.50$
$\phi_D{}^* = 0.025$		$V = 0.10$
$\phi_C = 0.750$		$\eta_L = 0.10$
$\phi_D = 0.250$		$\eta_{LL} = 0.23$
		$\eta_{LL}{}^* = 0.30$

6.3.1   Fixed nominal non-labor incomes

The fourth column in Table 6.5 shows that a reform from labor towards pollution taxes reduces aggregate private income by 0.06%, despite the increase in real after-tax wages by 0.55%. Pollution falls by 3.7%, while employment rises by 0.17%. The effects in the fourth column are the result of three separate effects. The first column shows that pollution taxes cause a tax burden effect, which is responsible for a reduction in wages, employment and private income. Second, by reducing the real value of transfers by 2.5% (see expression (T6.6)), the energy tax shifts the tax burden from high-skilled households towards low-skilled households, thereby stimulating employment by 0.4% and boosting aggregate private income by 0.31% (see the second column of Table 6.5). Finally, the energy tax raises the real value of the TFPA by 9.5% (see (T6.7)). This causes a reduction in employment and aggregate private income by, respectively, 0.14% and 0.10% (see the third column of Table 6.5). On balance, the tax shifting effect from labor to transfers does not dominate the other two effects, so that there is no utilitarian strong double dividend.

As discussed in section 6.2.1, this reform affects the distribution of income among the different households. In particular, we find that the high-skilled household, whose labor income share is relatively large, benefits from the environmental tax reform. Indeed, the higher real after-tax wage rate dominates the reduction in non-labor income, so that his private income increases by 0.3%. The low-skilled household, in contrast, suffers from a substantial decline in real transfers without benefitting much from a higher real wage rate. Accordingly, this household experiences a loss in private income by 1.17%.

6.3.2   Indexed real non-labor incomes

If the government maintains the income distribution by adjusting the level of transfers and the TFPA, the final column of Table 6.4 reveals that the pollution tax reduces employment by 0.05% and aggregate private income by 0.25%.[73] This implies that the high productive household is worse off compared to the tax reform in which non-labor incomes are not indexed. Furthermore, the decline in aggregate private income is larger than in the fourth column of Table 6.4. However, private income for the low productive household falls less compared to the previous experiment. Hence, if real non-labor incomes are indexed to real after-tax wages, the effects on employment and aggregate private income are less favorable than if non-labor incomes are not indexed. This trade-off between equity and efficiency was illustrated also by CPB (1993). Adopting the MIMIC

---

[73] The effect on private income in this chapter is defined in percentage changes. If it would be defined in terms of output, as is the case in the benchmark model, this value should be pre-multiplied by the output share of disposable household income (which equals 0.4 in chapter 3). By doing so, we find that the effect on private income in Table 6.5 is equivalent to the effect in the benchmark model. This is consistent with our theoretical findings in subsection 6.2.2.

model, CPB (1993) finds that an environmental tax reform typically reduces unemployment if that reform lowers the replacement rate. However, this comes at the expense of those collecting unemployment benefits.

Table 6.5: Effects according to the model with non-identical households of an increase in energy taxes that raises the consumer price of energy by 10%

	Non-labor incomes not indexed				Indexed non-labor incomes
	(1)	(2)	(3)	(4)	
Employment	-0.09	0.40	-0.14	0.17	-0.05
Pollution	-4.08	0.58	-0.20	-3.70	-4.02
Real after-tax wage	-0.38	1.42	-0.49	0.55	-0.25
Private income	-0.27	0.31	-0.10	-0.06	-0.25
Private income of high-skilled household[1]				0.30	-0.25
Private income of low-skilled household[2]			-1.17		-0.25

[1] High-skilled household is characterized by $\phi_L^i=.88$; $\phi_S^i=.11$; $\phi_D^{*i}=.01$
[2] Low-skilled household is characterized by $\phi_L^i=.16$ $\phi_S^i=.77$ $\phi_D^{*i}=.07$
(1) Tax burden effect
(2) Effect through lower real transfer income
(3) Effect through higher real TFPA
(4) Overall effect

6.4   Conclusions

This chapter modifies the benchmark model by introducing non-identical households. It illustrates the fundamental trade-off between aggregate private income, employment, equity, and the quality of the natural environment. Raising pollution taxes on consumption and using the revenues to cut distortionary labor taxes may yield a double dividend by raising aggregate private income, employment, and environmental quality. However, environmental taxes can alleviate labor-market distortions only by cutting the real value of non-labor incomes. Thus, the double dividend involves costs in terms of a less equitable income distribution. If the government imposes the restriction that real non-labor incomes are indexed to real after-tax wages, environmental policy will always reduce employment and aggregate private income and will thus fail to produce a double dividend. The government can employ various instruments to ensure maintenance of the income distribution. In particular, tax-free allowances and lump-sum transfers change the trade-off between equity and efficiency in similar ways.

7   Labor-market imperfections and the triple dividend

The competitive labor market in the benchmark model is not an appropriate characterization of the labor markets in Europe. Indeed, many European countries face a problem of high structural unemployment rates that are related to imperfections on the labor market. In order to gain insight into the effects of an environmental tax reform in those economies, this chapter introduces labor-market imperfections to our analysis.

The theory on imperfect labor markets provides alternative explanations for the phenomenon of involuntary unemployment (Layard et al., 1991). Efficiency wage models start from the idea that individual firms pay wages above the full-employment level in order to motivate workers (e.g. to keep them from shirking). Hence, involuntary unemployment is an essential disciplinary device for worker effort. In trade-union models, employers and unions bargain over wages. In particular, unions trade off the utility from higher wage incomes (relative to some fall-back position) with the level of employment among her members. In this way, unions may raise the wage rate above the market-clearing level. Finally, search models assume that a job match between a worker and an employee involves some rent, e.g. due to hiring costs. Through individual worker-firm bargaining, workers may reap part of these rents, thereby driving the wage rate above the market-clearing level.

This chapter develops a trade-union model to describe the process of wage formation. Accordingly, it investigates the opportunities for a double dividend in the presence of labor-market imperfections. The investigation does not start from the benchmark model of chapter 3. The reason is that the benchmark model is not an appropriate framework to introduce a wage-bargaining model, since it does not contain rents that can be bargained over. Such rents are necessary for wage negotiations to be meaningful. Therefore, we start from the model of section 4.4 which contains a fixed factor. The profits from this fixed factor form the rents in the negotiation process between unions and employers.

This chapter is related to earlier studies on the double dividend in the presence of labor-market imperfections. Bovenberg and van der Ploeg (1996, 1998b) explore the welfare effects of an environmental tax reform in the presence of fixed consumer wages. Van Hagen (1995) adopts a similar approach but incorporates heterogeneous agents and rationing of only low-skilled labor. These studies, however, do not explicitly describe the underlying labor-market imperfection that gives rise to involuntary unemployment.

Using an efficiency-wage framework, Schneider (1995) shows that an environmental tax reform may reduce involuntary unemployment. In particular, if the reform reduces utility in unemployment relative to utility on the job, wages may fall because lower wages suffice to maintain an optimal level of worker effort.

Bovenberg and van der Ploeg (1998a) adopt a model where labor-market imperfections originate in hiring costs, associated with a costly matching process between vacancies and the unemployed looking for a job. They incorporate several arguments in the fall-back position of the unemployed, including the utility from leisure, unemployment benefits and income from work in the informal sector. It turns out that including these different sources of utility is important for the employment double dividend. In particular, if sources of utility in unemployment are indexed to after-tax wages, an environmental tax reform has no impact on equilibrium unemployment. However, if some of these sources are not indexed to after-tax wages, an employment double dividend can be feasible.

Koskela and Schöb (1996), Holmlund and Kolm (1997) and Bayindir-Upmann and Raith (1997) investigate whether a double dividend is feasible in the presence of a trade-union model. Koskela and Schöb (1996) show that the employment effect of a reform from labor taxes towards environmental taxes on consumption depends on the indexation of unemployment benefits. In particular, employment rises if unemployment benefits are untaxed under the labor tax and not indexed to consumer prices. Intuitively, in that case an environmental tax reform reduces the fall-back position of the union, thereby causing downward wage pressure. In particular, the unemployed do not benefit from the lower labor taxes, but pay the higher environmental taxes on consumption. However, if unemployment benefits are taxed and indexed to consumer prices, employment declines because an environmental tax reform improves the bargaining position of the union. On the one hand, the unemployed do not pay the higher environmental tax because benefits are increased to compensate for the higher consumer prices. On the other hand, they benefit from the lower labor taxes. In that case, wages tend to rise, rather than fall. Holmlund and Kolm (1997) assume that utility in unemployment is indexed to the consumer wage, but nevertheless find that an environmental tax reform may boost employment. This is because the reform may relocate employment from sectors where unions exert relatively much bargaining power towards other sectors. For different forms of trade-union models, Bayindir-Upmann and Raith (1997) find that employment typically expands due to an ecological tax reform. The reason is that such a reform shifts the tax burden from labor towards capital. In particular, whereas both workers and capitalists bear the burden of pollution taxes, only workers benefit from reductions in the labor tax. As utility of the unemployed is determined by the value of leisure alone, the unemployed neither benefit from the lower labor tax nor suffer from a higher pollution tax. On balance, an ecological tax reform thus makes workers better off relative to the unemployed. This weakens the bargaining position of the union so that wage costs fall and employment expands. In the model of Bayindir-Upmann and Raith, however, this expansion in employment is not a double dividend because higher economic activity raises aggregate pollution. Hence, the government still faces a trade-off between protecting the environment and reducing unemployment.

This chapter differs from the above analyses by considering the competitive labor market as a benchmark case. Accordingly, we are able to investigate how imperfections themselves affect the opportunities for a double dividend. A second contribution of this chapter is that it helps us to better understand the variety in results from empirical models employed in the double-dividend debate. In particular, from the trade-union model we derive a wage equation that contains a number of variables that are affected by an environmental tax reform, including labor productivity, the labor tax rate and the unemployment rate. Different values for the elasticities in the wage equation explain a major part of the differences in results from empirical models. To illustrate, if the elasticity of the labor tax in the wage equation is smaller than unity, lower labor taxes not only raise after-tax income from work, but also reduce wage costs. This so-called real wage resistance may occur if income during unemployment consists partly of income that is not subject to the labor tax. Real wage resistance typically raises the opportunities for an employment double dividend. Our theoretical analysis, however, suggests that real wage resistance exists only if lower labor taxes improve the position of employees compared to those outside the labor force. Hence, it involves a trade-off between equity and efficiency. This underlying trade-off is generally not clear from empirical studies in which real wage resistance is responsible for the existence of a double dividend.

The rest of this chapter is organized as follows. Section 7.1 introduces a trade-union model to describe the process of wage formation. That section presents also the linearized model with an explicit wage equation and a measure for welfare. Section 7.2 demonstrates the welfare effects of an environmental tax reform for several special cases concerning the wage equation. These results are illustrated numerically in section 7.3. This section discusses also some empirical studies on the double dividend. Finally, section 7.4 concludes.

7.1   The model with an imperfect labor market

7.1.1   Structure of the model

This chapter adopts the so-called right-to-manage model to describe the process of wage formation. In particular, wages are determined by negotiations between a union and an employers organization. The wage outcome typically exceeds the market-clearing level. Hence, there is involuntary unemployment in equilibrium.

The wage-bargaining process

On one side of the wage-bargaining process, employers maximize profits (Π) with respect to the wage rate (W), and two inputs, labor (L) and a polluting input (E). Hence, the strategy of the employer's organization can be described as follows:

$$\underset{W, L, E}{Max} \quad \Pi \; = \; Y \, - \, (\, 1 + T_L \,)\, WL \, - \, (1 + T_E)E \qquad (7.1)$$

On the other side of the wage-bargaining process, trade union behavior is described by the maximization of the following utility function:

$$\underset{W}{Max} \quad \Omega \; = \; L\, [\, W \, - \, W^* \,]^\eta \qquad 0 < \eta < \infty \qquad (7.2)$$

where wages are defined in real terms and the consumer price index is taken as the numeraire. According to (7.2), the trade union cares about two issues. First is the real surplus their members get from work (W), relative to the opportunity costs of work, the so-called reservation wage (W^*) (or fall-back position). The parameter η indicates the relative importance of wages for the utility of the union. Second, utility depends on the level of employment among the union's members (L). We assume that unions do not optimize with respect to employment, but take the labor-demand strategy of firms as a dominant factor in the determination of employment.[74]

We adopt the following generalized Nash-bargaining maximization problem to describe the outcome of the negotiation process:

$$\underset{W}{Max} \quad \Pi^\alpha \, \Omega^{1-\alpha} \qquad (7.3)$$

---

[74] This implies that the wage outcome in a right-to-manage framework is not necessarily efficient. The wage outcome would be efficient if bargaining parties would negotiate also over employment. Such a framework is called the efficient bargaining model (see e.g. McDonald and Solow, 1981).

where α stands for the bargaining power of the employer. The first-order condition from this maximization problem yields the following solution of the bargaining process:

$$\frac{\alpha}{\Pi} \frac{\partial \Pi}{\partial W} + \frac{1-\alpha}{\Omega} \frac{\partial \Omega}{\partial W} = 0 \tag{7.4}$$

The partial derivatives on the LHS of (7.4) can be obtained from (7.1) and (7.2). By substituting these into (7.4), we derive the following expression for wages:

$$W = W^* + \eta W \left[\frac{\alpha}{1-\alpha} \frac{(1+T_L)WL}{\Pi} + \epsilon \right]^{-1} \tag{7.5}$$

where $\infty > \epsilon \equiv -(\partial L/\partial W)(W/L) > 0$ stands for the wage elasticity of labor demand. We assume that production (Y) features decreasing returns to scale with respect to its two inputs, L and E. This implies that profits (Π) are positive due to the presence of a fixed factor (see chapter 4). According to (7.5), positive profits imply that $W \geq W^*$. Indeed, the negotiated wage outcome equals the reservation wage of employees (W^*) plus some of the rents that can be bargained over.[75]

If the bargaining power of the employers organization is large (i.e. α is close to unity), employers succeed in keeping the wage level low, i.e. close to the reservation wage of employees. In that case, the rents flow towards the owners of the fixed factor in the form of profit incomes. This also holds if unions care little about wages compared to employment (so that η is small). Furthermore, wages are also lower if labor demand is elastic (i.e. is ϵ large). In this latter case, wage increases are expensive for the union, as they induce a strong adverse effect on employment. If unions care much about wages (i.e. η is large), if they exert relatively much bargaining power (i.e. $1-\alpha$ is close to unity), and if labor demand is inelastic (i.e. ϵ is small), unions may raise the wages of their

---

[75] If we start from the benchmark model of chapter 3, these rents would be zero. Hence, there would be no meaningful negotiation process, as the second term on the RHS of (7.5) would be zero. Rents may originate also in other sources. To illustrate, in Bovenberg and van der Ploeg (1998a) rents originate in hiring costs.

members. In that case, workers, rather than owners of the fixed factor, receive the rents.

The reservation wage

The reservation wage in (7.5) measures the opportunity costs from work. We adopt the following specification of the reservation wage:

$$W^* = (1-u)\,W + u\,B \qquad (7.6)$$

Work sharing implies that each member of the union spends an equivalent amount of time in employment and unemployment.[76] Therefore, the reservation wage in (7.6) strikes a balance between the utilities in case of employment (measured in monetary terms by the wage rate in the first term on the RHS of (7.6)) and unemployment (measured by the parameter B in the second term on the RHS of (7.6)). The parameter $u \equiv 1 - L/L_S$ stands for the time spent in unemployment of a representative member of the union. Utility during unemployment (B) is an important parameter for the wage outcome in wage-bargaining models. This utility may depend on several variables, including the level of unemployment benefits, untaxed income from the informal sector, the value of housekeeping activities, and the value of leisure. The next section pays more attention to the assumptions regarding B.

We define the reservation rate, R, as the ratio between the reservation wage and the market wage, i.e.:

$$R \equiv \frac{W^*}{W} = (1-u) + u\,\frac{B}{W} \qquad (7.7)$$

---

[76] The assumption of work sharing allows for an explicit derivation of three welfare dividends (see subsection 7.1.3). In our framework, this is attractive because it illustrates how labor-market imperfections modify the analysis of chapter 4. Without work sharing, we need to distinguish employed and unemployed households -- which complicates welfare analysis due to distributional issues (see chapter 6). Indeed, the welfare effects of an environmental tax reform could no longer be divided into three separate welfare dividends. Also Nielsen et al. (1995) adopt the assumption of work sharing in a wage-bargaining model to explore the opportunities for a double dividend.

The reservation rate measures the utility of a representative union member who is rationed in his labor supply (W^*), relative to his utility if there is unrationed labor supply (W). We assume that utility in case of involuntary unemployment is below the utility in employment i.e. $B < W$. This assumption ensures that unemployment is indeed involuntary, so that the reservation rate is smaller than unity. The reservation rate is used in the next sections to discuss the impact of policies on the relative bargaining position of the union in wage negotiations. In particular, a higher reservation rate indicates that unions have a strong bargaining position in wage negotiations. Hence, it is associated with a higher wage rate.

By substituting the reservation wage from (7.6) into (7.5), we derive the following expression for wages:

$$W = B + \frac{\eta}{u} W \left[\frac{\alpha}{1 - \alpha} \frac{(1 + T_L) W L}{\Pi} + \epsilon \right]^{-1} \tag{7.8}$$

Expression (7.8) reveals that the wage rate is determined by income during unemployment in addition to some of the rents that are bargained over between employers and the union.

7.1.2   Linearization

This chapter starts from the model of section 4.4 which includes a fixed factor. The model in relative changes is presented in Table 7.1. The table does not present fixed variables such as the fixed factor (H) and public consumption (G). In contrast to chapter 4, we assume that profit taxes are zero.

Table 7.1: Linearized model with fixed factor and non-competitive labor market

Domestic output	$\tilde{Y} = w_E \tilde{E} + w_L \tilde{L}$	(T7.1)
Polluting input demand	$\tilde{E} = -\epsilon_{EL}(\tilde{W} + \tilde{T}_L) - \epsilon_{EE}\tilde{T}_E$	(T7.2)

Labor demand $\qquad\qquad \tilde{L} = -\epsilon_{LL}(\tilde{W} + \tilde{T}_L) - \epsilon_{LE}\tilde{T}_E$ $\qquad\qquad$ (T7.3)

Profits $\qquad\qquad w_\Pi\tilde{\Pi} = -w_L(\tilde{W} + \tilde{T}_L) - w_E\tilde{T}_E$ $\qquad\qquad$ (T7.4)

Household budget $\qquad w_L(1 - \theta_L)(\tilde{L} + \tilde{W}) + w_\Pi\tilde{\Pi} = w_C\tilde{C}$ $\qquad$ (T7.5)

Labor supply $\qquad\qquad \tilde{L}_S = \eta_{LL}\tilde{W} - \eta_L\dfrac{w_\Pi}{w_C}\tilde{\Pi}$ $\qquad\qquad$ (T7.6)

Government budget $\qquad 0 = w_L\tilde{T}_L + w_E\tilde{T}_E + \theta_L w_L(\tilde{W} + \tilde{L}) + \theta_E w_E\tilde{E}$ (T7.7)

Wage rate $\qquad\qquad \tilde{W} = \beta\tilde{h} - \gamma\tilde{T}_L - \delta(\tilde{L}_S - \tilde{L})$ $\qquad\qquad$ (T7.8)

Value-added per worker $\qquad \tilde{h} = -\left(\dfrac{w_\Pi}{1 - w_E}\tilde{L} + \dfrac{w_E}{1 - w_E}\tilde{T}_E\right)$ $\qquad$ (T7.9)

Environmental quality $\qquad w_M\tilde{M} = \gamma_E w_E\tilde{E}$ $\qquad\qquad$ (T7.10)

Balance of payments $\qquad \tilde{Y} = w_C\tilde{C} + (1 - \theta_E)w_E\tilde{E}$ $\qquad\qquad$ (T7.11)

Parameters

ϵ_{IJ} $\quad$ = demand elasticity (conditional on the fixed factor) of factor I with respect to changes in the price of factor J, for $I,J = E,L$

η_{LL} $\quad$ = uncompensated labor-supply elasticity

η_L $\quad$ = income elasticity of labor supply

η_{LL}^{*} $\quad$ = compensated labor supply elasticity

β $\quad$ = elasticity that measures the effect of value-added per worker on wages

γ $\quad$ = elasticity that measures the effect of labor tax on wages

δ $\quad$ = elasticity that measures the effect of unemployment rate on wages

---

Factor demand
Expressions (T7.2) and (T7.3) represent the demand equations for, respectively, polluting inputs and labor. The demand elasticities, ϵ_{ij} for $i,j = L,E$, measure the impact of a change in producer prices on factor demands, for a given level of the fixed factor. In the factor demand equations, these elasticities are defined with a minus sign. Table 7.2 presents the elasticities for three separable production functions. It reveals that ϵ_{LL} and ϵ_{EE} are always positive. Hence, higher producer prices for labor and polluting inputs reduce the respective factor demands. The parameters ϵ_{LE} and ϵ_{EL} are positive if either labor or polluting inputs are separable in the production function from the other inputs. However, these parameters can be negative if the fixed factor is separable from other inputs, namely if the fixed factor is a relatively poor substitute for the two other inputs. In that case, a higher producer price of labor (polluting inputs) raises the demand for polluting inputs (labor). Intuitively, a change in relative producer prices induces a relatively large substitution effect between labor and polluting inputs. At the same time, the adverse output effect is small, as substitution with the fixed factor is difficult.

Value-added per worker
We define value-added per worker as the marginal product of labor (equal to the before-tax wage rate) plus profits per worker:

$$h \equiv (1+T_L)W + \frac{\Pi}{L} = \frac{Y}{L} - \frac{(1+T_E)E}{L} \tag{7.9}$$

In Table 7.1, expression (T7.9) shows the linearized expression for value-added per worker. It reveals that a higher pollution tax or higher employment, ceteris paribus, reduces the value-added per worker. The value-added per worker enters the linearized wage equation in (T7.8), which is discussed below.

Wage rate
The most important difference with the model of section 4.4 is that the equilibrium condition on the labor market is replaced by a linearized wage equation. In particular, (T7.8) expresses the relative change in the wage rate in

terms of changes in the value-added per worker (h), the labor tax (T_L) and the unemployment rate ($L_S - L$), where the elasticities are denoted by, respectively, ß, γ, and δ. These latter elasticities depend on the assumptions regarding the relative change in income during unemployment (B). In fact, by linearizing (7.8) and using (T7.4) to eliminate profits, we first derive the following linearized wage equation:

$$\tilde{W} = z_1 \tilde{B} + z_2 [\tilde{h} - \tilde{T}_L] - z_3 [\tilde{L}_S - \tilde{L}] \tag{7.10}$$

where

$$z_1 = \frac{w_H \Lambda B}{w_H \Lambda B + (1 - w_E)(W - B)} \qquad z_2 = \frac{(1 - w_E)(W - B)}{w_H \Lambda B + (1 - w_E)(W - B)}$$

$$z_3 = \frac{w_H \Lambda [W - B](1 - u)/u}{w_H \Lambda B + (1 - w_E)(W - B)}$$

where $\Lambda = 1 + \dfrac{1 - \alpha}{\alpha} \dfrac{w_H}{w_L} \epsilon$

so that $z_1 + z_2 = 1$. The first term on the RHS of (7.10) denotes the relative change in income during unemployment. In expression (T7.8), this term has been eliminated by substituting an expression for the relative change in B. In particular, we distinguish three different cases concerning the indexation of income during unemployment. For each of these indexation rules, we derive a different wage equation that can be expressed as in (T7.8), where ß, γ, and δ may be different. We consider the following rules:

1. Indexation to after-tax wages

The first indexation rule assumes that income during unemployment is indexed to after-tax wages, i.e. $B = aW$, where a is a constant parameter. This indexation rule would be relevant if income during unemployment would consist of unemployment benefits that are indexed to after-tax wages. The wage equation in (7.10) can then be written as follows:[77]

$$\tilde{W} = \tilde{h} - \tilde{T}_L - \frac{z_3}{z_2}[\tilde{L}_S - \tilde{L}]$$
(7.11)

where we used the fact that $z_1 + z_2 = 1$. Expression (7.11) is a special case of (T7.8). In particular, it reveals that, if income during unemployment is indexed to after-tax wages, the parameters β and γ in (T7.8) are equal to unity. Intuitively, changes in the value-added per worker or the labor tax leave the reservation rate in (7.7) unchanged, as they yield a similar impact on the utilities of employment and unemployment. Accordingly, the bargaining position of the union remains unchanged. Workers thus fully claim a higher value-added per worker in terms of a higher consumer wage. Moreover, they fully bear the incidence of higher labor taxes.

2. Indexation to value-added per worker
In the second indexation rule, income during unemployment is related to value-added per worker, i.e. $B = bh$, where b is a constant. The interpretation may be that value-added per worker is an indicator for the productivity in housekeeping activities or the underground economy. These activities might measure the opportunity costs of the time spent in unemployment. With this second indexation rule, the wage equation in (7.10) can be written as follows:[78]

$$\tilde{W} = \tilde{h} - z_2\tilde{T}_L - z_3[\tilde{L}_S - \tilde{L}]$$
(7.12)

---

[77] We do not include income during unemployment in the government budget constraint or the household budget constraint. The reason is that we allow for different interpretations of utility during unemployment, e.g. utility derived from housekeeping activities. Alternatively, one may include different forms of utility during unemployment explicitly, thereby also including them in the government and household budget constraints. This latter approach has been followed by Bovenberg and van der Ploeg (1998a).

[78] A similar expression would be derived if income during unemployment would be indexed to the before-tax wage.

If income during unemployment is indexed to the value-added per worker, expression (7.12) reveals that there exists real wage resistance. This is reflected by $\gamma < 1$ in expression (T7.8). Intuitively, higher labor taxes strengthen the bargaining position of the union by raising the reservation rate (see (7.7)). Indeed, whereas higher labor taxes reduce income from work in the formal sector, they leave untaxed, informal incomes in the underground economy or in household production unaffected. The stronger bargaining position of the union has an upward effect on the wage rate. Accordingly, the incidence of labor taxes is shifted onto employers.

3. No indexation

In the third case, income during unemployment is nominally fixed, i.e. $B = c$, where c is a constant. In that case, the relative change in B would be zero, so that (7.10) would read as follows:

$$\tilde{W} = z_2[\tilde{h} - \tilde{T}_L] - z_3[\tilde{L}_S - \tilde{L}] \tag{7.13}$$

Hence, if income during unemployment is constant, expression (7.13) shows that the parameters β and γ are both smaller than unity. Indeed, both a higher value-added per worker in the formal sector and a lower labor tax reduce the reservation rate in (7.7), as they raise the difference between the utilities from working and being unemployed. This weakens the bargaining position of unions in the wage-negotiation process. Accordingly, $\beta = \gamma < 1$.

Table 7.2: Price elasticities of factor demands in the model with fixed capital

	$Y = F[q(L,H),E]$	$Y = F[q(E,H),L]$	$Y = F[q(L,E),H]$
ϵ_{LL}	$\dfrac{(1 - w_E)\sigma_{LH}}{w_H}$	$\dfrac{w_H\sigma_L + w_L w_E\sigma_{EH}}{w_H(1 - w_L)}$	$\dfrac{w_L\sigma_H + w_E w_H\sigma_{LE}}{w_H(1 - w_H)}$
ϵ_{LE}	$\dfrac{w_E}{w_H}\sigma_{LH}$	$\dfrac{w_E}{w_H}\sigma_{EH}$	$\dfrac{w_E}{w_H}\dfrac{\sigma_H - w_H\sigma_{LE}}{1 - w_H}$

ϵ_{EE}	$\dfrac{w_H\sigma_E + \sigma_{LH}w_L w_E}{w_H(1-w_E)}$	$\dfrac{(1-w_L)\sigma_{EH}}{w_H}$	$\dfrac{w_E\sigma_H + w_L w_H\sigma_{LE}}{w_H(1-w_H)}$
ϵ_{EL}	$\dfrac{w_L}{w_H}\sigma_{LH}$	$\dfrac{w_L}{w_H}\sigma_{EH}$	$\dfrac{w_L}{w_H}\dfrac{\sigma_H - w_H\sigma_{LE}}{1-w_H}$
$\epsilon_{LL} - \epsilon_{EL}$	σ_{LH}	$\dfrac{\sigma_L - w_L\sigma_{EH}}{1-w_L}$	σ_{LE}
$\epsilon_{EE} - \epsilon_{LE}$	$\dfrac{\sigma_E - w_E\sigma_{LH}}{1-w_E}$	σ_{EH}	σ_{LE}
ϵ_D	$\dfrac{\sigma_{LH}\sigma_E}{w_H}$	$\dfrac{\sigma_L\sigma_{EH}}{w_H}$	$\dfrac{\sigma_H\sigma_{LE}}{w_H}$

$$\text{where } \epsilon_D \equiv \epsilon_{LL}\epsilon_{EE} - \epsilon_{LE}\epsilon_{EL} \quad \text{and} \quad w_L\epsilon_{LE} = w_E\epsilon_{EL}$$

For each indexation rule, an increase in the unemployment rate reduces the wage rate.[79] This is because a higher unemployment rate depresses the reservation rate in (7.7), as income from employment exceeds that from unemployment. Accordingly, a higher unemployment rate weakens the bargaining position of the union, so that wages fall. In (T7.8), this so-called wage curve is measured by the parameter δ. A small elasticity of the wage curve indicates strong rigidities on the labor market, i.e. wages do not respond much to the rate of unemployment. A large value for δ implies that market forces play an important role in the process of wage formation. Indeed, if the wage curve goes to infinity, the wage equation boils down to an equilibrium condition on the labor market.

---

[79] Since $B < W$, z_3 (and thus δ) is always non-negative.

7.1.3   Welfare

Work sharing implies that we can maintain the assumption of a representative household. Hence, we can perform welfare analysis by deriving the marginal excess burden from the utility function of the household model. In particular, each representative household faces a constrained optimization problem, where the number of labor hours is rationed by labor demand (L). Indeed, the wage outcome from the bargaining process between employers and the union exceeds the equilibrium wage level, so that the desired supply of labor exceeds the optimal demand for labor by firms. We define the so-called virtual wage, W^N, as the wage rate at which households would voluntarily choose $V = 1 - L$ as their optimal demand for leisure:

$$\frac{\partial u}{\partial V} = \lambda \, W^N \tag{7.14}$$

The demand for leisure according to (7.14) exceeds the desired demand for leisure that would be optimal for households at the going wage rate. Hence, the opportunity costs of desired leisure exceed the opportunity costs of leisure in the rationed optimum, i.e. $W > W^{N}$.[80]

Following the same procedure as in appendix 3B, and using the first-order condition in expression (7.14) instead of (3.5), we derive the following expression for the *MEB* in the presence of rationing in labor supply due to labor-market imperfections:

$$MEB = -[(1-\theta_L)w_L\tilde{W} + w_\Pi\tilde{\Pi}] - (1-\theta_L)(w_L - w_L^N)\tilde{L} - \frac{u_M}{\lambda}w_M\tilde{M} \tag{7.15}$$

blue	*pink*	*green*
dividend	*dividend*	*dividend*

---

[80] The household model measures utility in case of unemployment by the value of leisure, denoted by the virtual wage. This value depends on the concavity of the household utility function. The union, in contrast, measures utility in case of unemployment by the reservation wage, W^*. Although these measures for the utility during unemployment differ, they have a similar interpretation. In our welfare analysis, we adopt the virtual wage as an indicator for the value of unemployment.

where $(1-\theta_L)w_L^N = W^N L / P_Y Y$. In the presence of involuntary unemployment, the *MEB* in (7.15) contains one additional term compared to its equivalent in (4.20). In particular, the second term on the RHS of (7.15) reveals that a higher level of employment raises welfare if $w_L > w_L^N$, i.e. if the actual wage rate exceeds the virtual wage. As discussed above, this condition is met if there is involuntary unemployment. Intuitively, by reducing the demand for leisure, a higher level of employment improves welfare, as it relaxes the rationing constraint in the labor-supply decision of households. The magnitude of this welfare gain is measured by the gap between the actual wage and the virtual wage. In analogy to Bovenberg and van der Ploeg (1998b), we call the welfare gain associated with fewer non-tax distortions on the labor market the *pink dividend*.

The first and third term on the RHS of (7.15) are similar to those in expression (4.20). In particular, they denote the blue dividend and the green dividend, respectively. An environmental tax reform is said to yield a strong double dividend if it yields a blue dividend and a green dividend. We speak about a *triple dividend* if the strong double dividend is accompanied by a pink dividend.

7.2   Environmental tax reform

The model in Table 7.1 is solved along the lines of Appendix 4D. To clearly understand the implications of an environmental tax reform for welfare in the context of an imperfect labor market, this section considers several special cases concerning the parameters in the wage equation. In particular, subsection 7.2.1 starts with the case of fixed consumer wages, i.e. where all the elasticities in the wage equation are zero. Subsection 7.2.2. explores the effects of labor taxes and labor productivity on wages, thereby assuming that the wage curve is absent, i.e. $\delta = 0$. The case of a competitive labor market is discussed in subsection 7.2.3. Finally, subsection 7.2.4 demonstrates an intermediate case with a wage curve, in which the effects through the labor tax and labor productivity on wages are absent.

Table 7.3: Reduced forms in the model with $\delta = 0$

	$\tilde{T}_E$
$\Delta \tilde{L}$	$(1-\gamma)[\epsilon_{LL} - \epsilon_{EL} - \theta_E \epsilon_D] w_E + (1-\theta_L) w_L [\beta \dfrac{w_E}{1-w_E} \epsilon_{LL} - \gamma \epsilon_{LE}]$
$\Delta \tilde{E}$	$-(1-\gamma)[\epsilon_{EE} - \epsilon_{LE} - \theta_L \epsilon_D] w_L +$ $(1-\theta_L) w_L [\beta (\dfrac{w_H}{1-w_E} \epsilon_D + \dfrac{w_E}{1-w_E} \epsilon_{EL}) - \gamma \epsilon_{EE}]$
$\Delta (W + \tilde{T}_L)$	$-(1-\gamma)[w_E - \theta_L w_L \epsilon_{LE} - \theta_E w_E \epsilon_{EE}] -$ $\beta(1-\theta_L) w_L [\dfrac{w_E}{1-w_E} + \dfrac{w_H}{1-w_E} \epsilon_{LE}]$
$\Delta \tilde{W}$	$\gamma [w_E - \theta_L w_L \epsilon_{LE} - \theta_E w_E \epsilon_{EE}]$ $-\beta \dfrac{w_E}{1-w_E} [w_L - \theta_L w_L \epsilon_{LL} - \theta_E w_E \epsilon_{EL}] -$ $\beta \dfrac{w_H}{1-w_E} [(\epsilon_{LL} - \epsilon_{EL} - \theta_E \epsilon_D)] w_E$
$\Delta w_\Pi \tilde{\Pi}$	$(1-\gamma)[\theta_L (\epsilon_{LL} - \epsilon_{EL}) - \theta_E (\epsilon_{EE} - \epsilon_{LE})] w_E w_L$ $+ (1-\theta_L) w_L [\beta (\dfrac{w_L}{1-w_E} + \dfrac{w_H}{1-w_E} (\epsilon_{LL} - \epsilon_{EL})) - \gamma] w_E$

$\Delta \tilde{B}$	$(1-\gamma)[\theta_L(\epsilon_{LL}-\epsilon_{EL}) - \theta_E(\epsilon_{EE}-\epsilon_{LE})]w_E w_L$
	$+ \beta(1-\theta_L)w_L \dfrac{w_E}{1-w_E}[\theta_L w_L \epsilon_{LL} + \theta_E w_E(\dfrac{w_H}{w_E}\epsilon_D + \epsilon_{EL})]$
	$- \gamma(1-\theta_L)w_L[\theta_L w_L \epsilon_{LE} + \theta_E w_E \epsilon_{EE}]$

$$\Delta = (1-\gamma)[w_L - \theta_L w_L \epsilon_{LL} - \theta_E w_E \epsilon_{EL}] + (1-\theta_L)w_L[\gamma - \beta \frac{w_H}{1-w_E}\epsilon_{LL}]$$

7.2.1   Fixed consumer wages[81]

Table 7.3 gives the reduced-forms when $\delta = 0$ for, respectively, employment, polluting input demand, producer wages, consumer wages, profits and aggregate private income. In this subsection, we consider the extreme situation of a fixed consumer wage that does not change due to policy shocks. In Table 7.3, this situation is characterized by $\beta = \gamma = 0$.

Pink welfare
The first term between square brackets in the first row of Table 7.3 (with $\beta = \gamma = 0$) reveals that an environmental tax reform affects employment through two channels. First, the reform induces a substitution effect from polluting inputs towards labor and an ambiguous output effect. On balance, employment expands if the value of the own-price elasticity of labor demand exceeds that of the cross-price elasticity, i.e. $\epsilon_{LL} - \epsilon_{EL} > 0$. According to Table 7.2, this condition is typically met. In particular, employment falls only if labor is separable from other inputs and, compared to polluting inputs, is a much poorer substitute for

---

[81] This case corresponds to Bovenberg and van der Ploeg (1998b). Indeed, the reduced-form equations in Table 7.1 are equivalent to theirs if $\beta = \gamma = \delta = 0$.

the fixed factor (i.e. $\sigma_L < w_L \sigma_{EH}$). In that case, an environmental tax reform causes a large adverse output effect because it shifts the tax burden from the factor that is inelastic in demand (labor) towards the factor that is elastic in demand (the polluting input). We refer to this as the exceptional case. In the normal case, employment expands.

The second channel through which an environmental tax reform affects employment is related to the effect on the tax base (see the second term between the first square brackets). In particular, a smaller base of the pollution tax reduces public revenues if $\theta_E > 0$. Accordingly, the government requires higher labor taxes to meet its revenue requirement. This raises the producer wage and reduces employment.

On balance, an expansion of employment is possible only if we are in the normal case and the second effect is small compared to the first effect. This requires that the initial pollution tax is sufficiently small. Indeed, from the reduced form in Table 7.3, we can derive the maximum pollution tax, θ_E, for which a marginal change in the tax structure between labor and pollution does not reduce employment. This value of θ_E thus optimizes pink welfare. For three separable production functions, the pink welfare optimum is presented in the first row of Table 7.4. We see that the pink optimum for θ_E is larger if polluting inputs are a poor substitute for the fixed factor, i.e if σ_E, σ_{EH} and σ_H are small. Intuitively, with poor substitution between E and H, polluting inputs are inelastic in demand. Hence, pollution taxes do not erode the base of the pollution tax by much. This renders the adverse effects on employment also small. If the initial equilibrium features a pollution tax below the pink welfare optimum, an environmental tax reform boosts employment. However, employment falls if the initial pollution tax is higher than the pink welfare optimum.

Green welfare
The effect of an environmental tax reform on pollution operates through the same channels as the effect on employment. In particular, the term between the first square brackets in the second row of Table 7.3 with $\beta = \gamma = 0$ reveals that substitution and output effects are captured by the term $\epsilon_{EE} - \epsilon_{LE}$. According to Table 7.2, polluting input demand typically falls through this first channel, apart

from the exceptional case that polluting inputs are separable from other inputs in production and are a relatively poor substitute for the fixed factor, i.e. $\sigma_E <$ $w_E \sigma_{LH}$. Second, polluting input demand rises through a broadening of the labor-tax base if $\theta_L > 0$.

From the reduced form for pollution, we derive the value of θ_L that maximizes green welfare. If the initial labor tax is below this green welfare optimum, an environmental tax reform reduces pollution and thus yields a green dividend. However, if the initial labor tax is larger, an environmental tax reform raises pollution so that the green dividend fails. The second row of Table 7.4 presents the optimal green θ_L for three separable production functions. It reveals that the optimal green labor tax is small if labor is a good substitute for the fixed factor. Intuitively, with easy substitution, lower labor taxes cause a large positive output effect, thereby raising pollution as well.[82]

Blue welfare
With fixed consumer wages, the incidence of labor and pollution taxes bears on profit incomes alone. If we start from an arbitrary equilibrium, the government may improve the efficiency with which these profit incomes are taxed by changing the tax mix between labor and the polluting inputs.[83] In particular, the government raises private incomes if labor is overtaxed in the initial equilibrium (compared to pollution). The term between the first square brackets in the fifth row of Table 7.3 reveals under which two conditions this is the case. First, labor taxes should be large compared to taxes on polluting inputs (i.e. θ_L large compared to θ_E). In that case, the broadening of the labor tax base boosts tax revenues substantially while the eroding base of the pollution tax has only a meager impact on overall tax revenues. Second, labor should be elastic in demand compared to pollution (i.e. $\epsilon_{LL} - \epsilon_{EL} > \epsilon_{EE} - \epsilon_{LE}$). Indeed, in accordance

---

[82] Other studies have also found that environmental taxes may harm environmental quality; see e.g. Bayindir-Upmann and Raith (1997), Brunello (1996), Schöb (1996) and chapter 4 of this book.

[83] Note that a profit tax would be a more efficient instrument to tax away the rents from the fixed factor; see chapter 4. This chapter assumes that this instrument is not available.

with the Ramsey tax principle, one should tax the input that is relatively inelastic in demand. If these two conditions are met, an environmental tax reform makes the tax system more efficient from a revenue-raising point of view, thereby raising private incomes.

From the reduced form for private incomes in Table 7.3, we derive the optimal tax structure from a blue welfare point of view. For three separable production functions, this tax structure is illustrated in the last row of Table 7.4. We find that the government should levy the highest tax on the factor that is a relative complement to the fixed factor. Indeed, poor substitution with the fixed factor implies that an input is relatively inelastic in demand. The inverse elasticity rule from the theory of optimal taxation suggests that this factor should feature the highest tax. If polluting inputs and labor are equally good substitutes for the fixed factor, Table 7.4 indicates that the optimal blue tax structure should not differentiate between labor and pollution taxes.

Table 7.4: Optimal tax structure from a blue, pink or green point of view, for three separable production functions.

	General case	$Y=F[E,g(L,H)]$	$Y=F[L,g(H,E)]$	$Y=F[H,g(L,E)]$
Pink welfare optimum	$\theta_E = \dfrac{\epsilon_{LL} - \epsilon_{EL}}{\epsilon_D}$	$\dfrac{w_H}{\sigma_E}$	$\dfrac{w_H}{1-w_L}\left(\dfrac{1}{\sigma_{EH}} - \dfrac{w_L}{\sigma_L}\right)$	$\dfrac{w_H}{\sigma_H}$
Green welfare optimum	$\theta_L = \dfrac{\epsilon_{EE} - \epsilon_{LE}}{\epsilon_D}$	$\dfrac{w_H}{1-w_E}\left(\dfrac{1}{\sigma_{LH}} - \dfrac{w_E}{\sigma_E}\right)$	$\dfrac{w_H}{\sigma_L}$	$\dfrac{w_H}{\sigma_H}$
Blue welfare optimum	$\dfrac{\theta_E}{\theta_L} = \dfrac{\epsilon_{LL} - \epsilon_{EL}}{\epsilon_{EE} - \epsilon_{LE}}$	$1 - \dfrac{\sigma_E - \sigma_{LH}}{\sigma_E - w_E \sigma_{LH}}$	$1 - \dfrac{\sigma_{EH} - \sigma_L}{(1-w_L)\sigma_{EH}}$	1

Overall welfare

The *MEB* in (7.15) reveals that, by changing the tax structure between labor and pollution, the government may affect welfare through three channels: private income (blue welfare), employment (pink welfare) and a cleaner environment (green welfare). Our tax reform analysis suggests that an environmental tax reform may simultaneously raise these three welfare components, thereby yielding a so-called triple dividend. In particular, a blue and a pink dividend can be obtained if the initial equilibrium is characterized by small pollution taxes, while polluting inputs should be a relative complement to the fixed factor (see the conditions in the first and third rows of Table 7.4). Furthermore, a green dividend is obtained if labor taxes are sufficiently small and if labor and the fixed factor cannot be substituted too easily (see the second row of Table 7.4).[84]

7.2.2   Effects of labor taxes and labor productivity

This subsection discusses the implications of an environmental tax reform through two additional channels compared to the previous subsection, namely the effects of labor taxes and labor productivity on wage formation, i.e. γ and β, respectively. To focus on these two channels, we maintain the assumption that the unemployment rate has no impact on wages, i.e. $\delta = 0$. The relevant reduced-form equations are presented in Table 7.3.

---

[84] The conditions on the production function for the green and pink dividends to hold might be conflicting. In particular, if either labor or the polluting input is separable from the other inputs, the condition that the polluting input is a relative complement to the fixed factor (which makes a pink dividend more likely) is not compatible with the condition that labor and the fixed factor are relative complements (which makes a green dividend more likely). Hence, with these production functions, a triple dividend is possible only under relatively strict conditions regarding the initial tax rates. If the fixed factor is separable from the other inputs, labor and the polluting input are equally good substitutes for the fixed factor. Hence, the conditions on the production function that labor and polluting inputs should be relative complements with the fixed factor are compatible with each other.

Indexation to after-tax wages

The previous section assumes fixed consumer wages. In that case, on the margin, both labor taxes and pollution taxes are entirely borne by employers. If income during unemployment is indexed to after-tax wages, however, labor and pollution taxes are borne by workers, rather than by employers. In particular, with $\gamma = \beta = 1$, lower labor taxes have no effect on the producer wage, but only raise the consumer wage. Furthermore, a lower value-added per worker, induced by the pollution tax, causes a proportional reduction in the wage rate (see (T7.8)).

The lower producer wage and the higher producer price of polluting inputs may affect employment. In particular, the solution in the first row of Table 7.3 with $\gamma = \beta = 1$ suggests that the effect on employment depends on the production function. Intuitively, whether or not labor demand expands due to an environmental tax reform depends on what happens to the producer wage relative to the marginal product of labor. Since the effect on the producer wage is determined by average value-added per worker, these effects do not necessarily coincide. In particular, if the marginal product of labor rises (falls) relative to the producer wage, an environmental tax reform increases (decreases) labor demand.

For three separable production functions, the first row of Table 7.5 presents under what conditions on the production function does an environmental tax reform yield a positive effect on employment. It reveals that employment remains unchanged if labor and the fixed factor are equally good substitutes for the polluting input (i.e. if the polluting input is separable or if, in all other cases, the substitution elasticities are equivalent, i.e. $\sigma_L = \sigma_{EH}$ or $\sigma_{LE} = \sigma_H$). Intuitively, in that case, the reduction in polluting inputs has an equivalent effect on the marginal productivities of labor and the fixed factor. Accordingly, the reduction in the producer wage -- which depends on both the marginal product of labor and the marginal product of the fixed factor -- is equal to the reduction in the marginal product of labor.

If, however, compared to the fixed factor, labor is a better substitute for the polluting input (i.e. if labor is separable and σ_L is large compared to σ_{EH}, or if the fixed factor is separable and σ_{LE} is large compared to σ_H), the pollution tax exerts a relatively large effect on the marginal product of the fixed factor (and thus on profits), and a relatively small effect on the marginal product of labor. Since wages respond to both the lower marginal product of labor and the lower marginal product of the fixed factor (see (7.8)), the relatively large decline in the marginal product of the fixed factor renders the effect on the producer wage larger than the decline in the marginal product of labor. Accordingly, producers find it attractive to raise their demand for labor. For similar reasons, an environmental tax reform reduces employment if the fixed factor and the polluting input are relatively good substitutes (i.e. if labor is separable and $\sigma_L <$ σ_{EH}, or if the fixed factor is separable and $\sigma_H > \sigma_{LE}$).

The lower part of Table 7.5 shows the implications of an environmental tax reform for the green dividend if $\gamma = \beta = 1$. It reveals that pollution typically falls on account of an environmental tax reform. The blue dividend depends on how changes in pollution and employment affect the respective tax bases. In particular, a positive blue dividend is feasible only if employment expands on account of an environmental tax reform and if labor is taxed in the initial equilibrium. This condition ensures that private incomes increase on account of a broader base of the labor tax. At the same time, the initial pollution tax should be low, especially if pollution falls substantially. If these conditions are not met, then the blue dividend fails. A triple dividend is thus obtained only if the pink dividend is positive and large, while the green dividend is positive but small.

Alternative indexation rules
If income during unemployment is indexed to value-added per worker, section 7.1.2 reveals that $\beta = 1$ and $\gamma < 1$ (i.e. there exists real wage resistance). According to Table 7.3, real wage resistance raises the opportunities for a pink dividend. Intuitively, in the presence of real wage resistance, lower labor taxes weaken the bargaining position of the union, since they reduce the income from unemployment relative to employment (i.e. they reduce the reservation rate in (7.7)). Accordingly, the producer wage per unit of output drops. Indeed, employers also reap part of the lower labor tax so that employment expands.

If income during unemployment is constant, section 7.1.2 shows that $\beta = \gamma$ < 1. According to the reduced forms in Table 7.3, environmental taxes yield less favorable effects on employment if $\beta < 1$. Intuitively, not only the lower labor tax, but also the reduction in labor productivity, affects the reservation rate. Indeed, as the income from unemployment is not indexed to productivity in the formal sector, a lower value-added per worker in the formal sector strengthens the bargaining position of the union, thereby driving up wages. Accordingly, the employees shift part of the burden of lower productivity onto the employers in terms of a higher producer wage. This reduces employment.

In general, the effect on employment depends importantly on the magnitude of ß relative to γ. Indeed, the effect on employment is most likely to be positive if ß = 1 and γ < 1. Most models of imperfect labor markets indeed assume that ß = 1, as this ensures that the equilibrium rate of unemployment remains constant in a growing economy. In a number of models, this implies also that γ = 1. To illustrate, by assuming that the unemployed collect incomes that are indexed to after-tax wages, Layard et al. (1991) argue that labor taxes are borne entirely by workers in the long run. In our model, however, γ may be smaller than unity if utility during unemployment is determined by untaxed income, such as housekeeping activities or income from underground activities. Whether real wage resistance is relevant is an empirical matter that will be discussed in section 7.3.

Discussion

The results in this section differ partly from the findings by Bovenberg and van der Ploeg (1998a). Using a model of an imperfect labor market due to hiring costs, they incorporate several determinants in the reservation wage, including unemployment benefits and pay from the informal sector. Accordingly, Bovenberg and van der Ploeg derive a number of conclusions related to the double dividend. First, if the reservation wage is indexed to the consumer wage, environmental taxes and labor taxes do not affect unemployment. If the reservation wage is indexed to after-tax wages, i.e. γ = ß = 1, however, we find that employment remains unchanged only if labor and the fixed factor are equally good substitutes for the polluting input. Hence, in contrast to the model of Bovenberg and van der Ploeg, the structure of the production function matters for the effect on employment in our model. The reason for this difference is that the rents that are bargained over in the wage negotiations do not originate in hiring costs (which are independent of the structure of the production function), as in Bovenberg and van der Ploeg, but in a fixed factor.

In our model, the production function thus matters for the effect on employment, even if income during unemployment is indexed to after-tax wages. Using a model where profits originate in mark-up pricing on the goods market, Koskela et al. (1998) derive a similar conclusion. In particular, they show that employment rises only on account of an environmental tax reform if labor and polluting inputs are sufficiently good substitutes for each other.

Table 7.5: Reduced form for employment and pollution if $\gamma = \beta = 1$, for three separable production functions

General case	$F[E,g(L,H)]$	$F[L,g(H,E)]$	$F[H,g(L,E)]$
Employment			
$\dfrac{w_E \epsilon_{LL} - (1-w_E)\epsilon_{LE}}{1 - w_E - w_H \epsilon_{LL}}$	0	$\dfrac{w_E(\sigma_L - \sigma_{EH})}{w_H(1-\sigma_L) + w_L w_E(1-\sigma_{EH})}$	$\dfrac{w_E(\sigma_{LE} - \sigma_H)}{w_L(1-\sigma_H) + w_E w_H(1-\sigma_{LE})}$
Pollution			
$\dfrac{w_H \epsilon_D + w_E \epsilon_{EL} - (1-w_E)\epsilon_{EE}}{1 - w_E - w_H \epsilon_{LL}}$	$-\dfrac{\sigma_E}{1-w_E}$	$-\dfrac{(1-w_L)(1-\sigma_L)\sigma_{EH}}{w_H(1-\sigma_L) + w_L w_E(1-\sigma_{EH})}$	$-\dfrac{w_L\sigma_{LE}(1-\sigma_H) + w_E\sigma_H(1-\sigma_{LE})}{w_L(1-\sigma_H) + w_E w_H(1-\sigma_{LE})}$

A second result of Bovenberg and van der Ploeg is that labor taxes depress employment, while pollution taxes leave employment unchanged if the reservation wage is determined by factors that are indexed to labor productivity, such as wages in the underground economy. If, in our model, income during unemployment is indexed to value-added per worker (i.e. $\gamma < 1$ and $\beta = 1$), this provides more potential for positive effects on employment. Hence, this result is consistent with Bovenberg and van der Ploeg.

7.2.3   Perfectly competitive labor market

Table 7.6 shows the reduced-form equations if $\gamma = \beta = 0$ and $\delta > 0$. This section discusses the specific case if $\delta \to \infty$. In that case, the wage equation in (T7.8) boils down to an equilibrium condition on the labor market. Hence, the effects are equivalent to those in section 4.6, which extensively discusses the effects of an environmental tax reform. Here, we briefly repeat the explanation of these results and focus on how these effects differ from those in subsection 7.2.1 and the benchmark model.[85]

In the presence of a competitive labor market, an environmental tax reform affects private welfare through two channels. The first term in the reduced form for private welfare in Table 7.6 denotes the efficiency with which the government taxes away profit incomes. This term closely resembles that of section 7.2.1 and contains two components: first, an effect induced by higher pollution taxes and, second, an effect induced by lower labor taxes. The first effect, through higher pollution taxes, reduces the demand for polluting inputs if $\epsilon_{EE} - \epsilon_{LE} > 0$. This erodes the base of the pollution tax if $\theta_E > 0$, thereby reducing the scope for a lower labor tax. Accordingly, blue welfare falls. The second effect, through lower labor taxes, stimulates labor demand if $\epsilon_{LL} - \epsilon_{EL} > 0$. In the competitive labor market, higher labor demand raises employment only if labor supply increases as well. In particular, in order to maintain the labor-market equilibrium, the market wage rate needs to rise so that labor supply is encouraged (provided that the compensated labor supply elasticity is positive, i.e. $\eta_{LL}^* > 0$). Hence, the higher level of employment is associated with a shift in the tax burden from profits to labor incomes. The expansion of employment broadens the base of the labor tax, thereby increasing tax revenues if $\theta_L > 0$. For a given public revenue requirement, this allows for lower marginal tax rates. Accordingly, blue welfare increases.

---

[85] The effects of labor taxes and labor productivity on wages (i.e. through β and γ) are irrelevant if the labor market is in equilibrium. Indeed, these elements are 'overruled' by competitive forces that determine the effects on the equilibrium wage rate.

The second channel through which an environmental tax reform affects private income is captured by the second term in the reduced form for private income in Table 7.6. This term reveals that private income typically falls due to an environmental tax reform if the initial pollution tax rate is positive. Intuitively, the pollution tax bears also directly on labor incomes. To tax labor incomes, it is typically more efficient to use direct taxes on labor, rather than a tax on polluting inputs (see chapter 3). Hence, a shift from labor to pollution taxes reduces the efficiency with which labor incomes are taxed, which is reflected in a reduction in private income.

By expanding employment and reducing the demand for polluting inputs, a swap of pollution taxes for labor taxes may boost non-environmental welfare only if the marginal welfare gain of additional employment dominates the marginal loss in non-environmental welfare due to a drop in polluting inputs. In addition, this welfare gain (represented by a positive first term in the reduced form for private welfare) should dominate the higher tax burden on labor (represented by the second term in the reduced form for private welfare). In that case, compared to polluting inputs, labor is overtaxed initially so that substituting pollution taxes for labor taxes shifts the tax system towards its non-environmental optimum.

The model of subsection 7.2.1 is characterized by the following two assumptions:

I.   The production function includes a fixed factor;
II.   Consumer wages are fixed, i.e. $\gamma = \beta = \delta = 0$.[86]

The first assumption implies that the incidence of labor and pollution taxes is shared across labor incomes and profit incomes. The second one indicates that workers do not bear the economic burden of tax changes. Hence, tax incidence is borne by profits alone. Moreover, only producer wages (not consumer wages)

---

[86] The case with $\delta = 0$ can also be interpreted as if the elasticity of labor supply goes to infinity. A positive value of δ thus requires a less-than-infinite elasticity of labor supply.

matter for the effect on employment, as this is determined at the short end of the labor market, i.e. by labor demand. Indeed, labor supply is irrelevant for employment in this model.

The benchmark model differs from that of subsection 7.2.1 in two respects:

1.   The production function does not contain a fixed factor;
2.   The labor market is characterized by perfect competition, i.e. $\delta \rightarrow \infty$.

The first assumption of the benchmark model indicates all taxes are ultimately borne by labor. The second assumption implies that not only producer wages (which determine labor demand), but also consumer wages (which affect labor supply) matter for employment.

This section forms an intermediate case between the benchmark model and the model of subsection 7.2.1. In particular, the model is characterized by the following two assumptions:[87]

1.   The production function contains a fixed factor (as in subsection 7.2.1);
2.   The labor market is characterized by perfect competition (as in the benchmark model).

Hence, in this subsection, the tax burden can be borne by both labor income and the income from the fixed factor. Furthermore, apart from labor demand, also labor supply matters for employment. The analysis of this section indeed contains a mixture of the results from the benchmark model and the results from subsection 7.2.1. On the one hand, an environmental tax reform reduces private incomes, as it makes the tax system less efficient as an instrument to tax away labor incomes. This so-called tax burden effect is similar to the result from the benchmark model. On the other hand, an environmental tax reform may improve

---

[87] The other intermediate case, i.e. without a fixed factor and with an imperfect labor market, cannot be explored because then we have no rents to be bargained over; see also footnote 83.

the efficiency with which profit incomes are taxed away. The conditions for this second effect to be positive are similar to those in section 7.2.1. In addition, they require that labor is supplied elastically.

7.2.4   Wage curve

If $0 < \delta < \infty$ and $\beta = \gamma = 0$, we are somewhere between the models of subsection 7.2.1 (without competitive forces) and subsection 7.2.3 (with a competitive labor market). In particular, subsection 7.2.1 reveals that an environmental tax reform may raise labor demand under particular conditions of the production function and initial tax rates. Through the wage curve, such an expansion in labor demand exerts upward wage pressure. This, on the one hand, mitigates the positive effect on labor demand. On the other hand, it stimulates labor supply. Accordingly, the effects on labor supply and demand converge. As employment is determined on the short end of the labor market as long as the labor market is not fully competitive, i.e. if $\delta < \infty$, the smaller effect on labor demand compels strong competitive forces to reduce the quantitative effect on employment (see Table 7.6). If δ moves close to infinity, the effects on labor demand and supply are equivalent and we are in the model of subsection 7.2.3.

Table 7.6: Reduced forms in case of a non-competitive labor market, $\beta = \gamma = 0$

	$\tilde{T}_E$
$\Delta \tilde{L}$	$(1 + \delta \eta_{LL}^*)[\epsilon_{LL} - \epsilon_{EL} - \theta_E \epsilon_D] w_E + \delta (1 - \theta_L) w_L \dfrac{\eta_L}{w_C} \theta_E w_E \epsilon_D$
$\Delta \tilde{E}$	$-(1 + \delta \eta_{LL}^*)[\epsilon_{EE} - \epsilon_{LE} - \theta_L \epsilon_D] w_L -$ $\delta (1 - \theta_L) w_L [1 + \theta_L w_L \dfrac{\eta_L}{w_C}] \epsilon_D$
$\Delta \tilde{W}$	$\delta [(\epsilon_{LL} - \epsilon_{EL} - \theta_E \epsilon_D) + \eta_L \dfrac{w_L}{w_C}(\theta_L(\epsilon_{LL} - \epsilon_{EL}) - \theta_E(\epsilon_{EE} - \epsilon_{LE})]w_E$

$$\Delta\, w_{\Pi}\, \tilde{\Pi} \qquad \begin{array}{c} (1+\delta\eta_{LL})\,[\,\theta_L\,(\epsilon_{LL}-\epsilon_{EL}) \;-\; \theta_E(\epsilon_{EE}-\epsilon_{LE})\,]\,w_E\,w_L \\[2mm] -\;\delta\,(1-\theta_L)w_L(\epsilon_{LL}-\epsilon_{EL})\,w_E \end{array}$$

$$\Delta\, \tilde{B} \qquad \begin{array}{c} (1+\delta\eta_{LL}^{*})\,[\,\theta_L\,(\epsilon_{LL}-\epsilon_{EL}) \;-\; \theta_E(\epsilon_{EE}-\epsilon_{LE})\,]\,w_E\,w_L \\[2mm] -\;\delta\,(1-\theta_L)\,w_L\,\theta_E w_E \epsilon_D \end{array}$$

$$\Delta \;=\; (1+\delta\eta_{LL})\,[\,w_L \,-\, \theta_L w_L\,\epsilon_{LL} \,-\, \theta_E w_E \epsilon_{EL}\,] \;+\; \delta\,(1-\theta_L)\,w_L[\,\epsilon_{LL} \,+\, \frac{w_L}{w_C}\,\eta_L\,]$$

$$\eta_{LL}^{*} \;=\; \eta_{LL} \,+\, \eta_L \frac{(1-\theta_L)w_L}{w_C}$$

7.3   Numerical simulations

This section presents some numerical simulations with the model of Table 7.1. The calibration of most of the parameters and elasticities is similar to that in section 4.7 (see Table 4.9). Table 7.7 presents the simulation results of an energy tax that raises the producer price of energy by 10%. The revenues are recycled through lower labor taxes. In Table 7.7, we take different values for the elasticities in the wage equation.[88]

7.3.1   Fixed consumer wages

The first column of Table 7.7 presents the simulation results if the consumer wage is fixed (i.e. $\beta = \gamma = 0$). In this case, we find that the reform from labor taxes towards energy taxes boosts employment by 1.3% and reduces pollution

---

[88] Note that the assumption of a fixed capital supply is important for these outcomes. Indeed, it suggests that these results should be interpreted as short-run effects, as capital is more mobile in the long run. If capital would be mobile, a reform from labor to energy taxes is less likely to raise employment and private income (see chapter 4).

by 5.12%. The rise in employment is valuable in terms of additional tax revenues because the initial labor tax is large. In contrast, the drop in pollution is cheap because of a small initial pollution tax. Accordingly, private incomes rise by 0.4%. Hence, labor seems overtaxed compared to pollution in the initial equilibrium. An environmental tax reform produces a triple dividend.

7.3.2   Indexation to after-tax wages

If income during unemployment is indexed to after-tax wages (i.e. $\beta = \gamma = 1$), section 7.2.2 finds that the effect on employment depends on the parameters in the production function. In particular, employment expands only if, compared to the fixed factor, labor is a better substitute for polluting inputs. Empirical evidence indeed suggests that this is the case if polluting inputs are interpreted as energy inputs. Hence, with our parameterization based on this evidence, we find that a reform from labor to energy taxes raises employment. The incidence of the pollution tax is borne by both profits and wages. Workers are compensated for the lower wage through the reduction in the labor tax. This positive effect on wages is reinforced through the government budget constraint. In particular, the expansion of employment broadens the tax base and thus raises tax revenues endogenously. With fixed government outlays, this allows for a further reduction in the labor tax, which fully accrues to workers in terms of a higher consumer wage. On balance, the positive effect on wages dominates the negative effect on profits, so that private incomes increase by 0.14%. Hence, with $\beta = \gamma = 1$, an energy tax reform yields a triple dividend. In contrast to the previous case with fixed consumer wages, however, this triple dividend is associated with a shift in income from profits towards wages.

7.3.3   Real wage resistance

If income during unemployment is indexed to value-added per worker (i.e. $\beta = 1$, $\gamma < 1$), section 7.2.2 suggests that an environmental tax reform is more likely to raise employment than with indexation to after-tax wages. In particular, if we take $\beta = 1$ and $\gamma = 0.5$, the third column of Table 7.7 shows that wages fall while profit incomes rise. The reduction in wage costs raises employment by 1.98%. Private incomes increase by 0.64%. These positive effects on employment and private income are larger than without real wage resistance ($\gamma = 1$). However, the triple dividend is due to large shifts in the income distribution. First, Table 7.7 reveals that consumer wages fall by 0.56%, while profits increase by 4.18%. Second, real wage resistance exists only if an environmental tax reform weakens the bargaining position of the union by widening income differentials between the employed and the unemployed.[89]

Empirical evidence on wage formation suggests that real wage resistance is important. In particular, Knoester and van der Windt (1987) find that the tax wedge has permanent effects on wages in ten OECD countries. More recently, Graafland and Huizinga (1997) estimate a non-linear wage equation for the Netherlands and find a long-run wage elasticity of the tax burden of 0.5 (on average over the sample period).[90] According to Layard et al. (1991), real wage resistance is unlikely to be permanent, however. Indeed, they argue that the empirical findings that support real wage resistance may be misleading, since these studies do not distinguish between permanent and temporary effects. Indeed, Layard et al. argue that real wage resistance is often long lasting, but not permanent. Brunello (1996) confirms this claim. In particular, estimating wage equations for all EU countries, he finds that short-term real wage resistance is important for some countries, especially for the Netherlands, but in the long term it is irrelevant. Our theoretical model shows, however, that real wage resistance

---

[89] Work sharing implies that these latter distributional concerns are irrelevant in our model. However, real wage resistance would raise an important distributional issue if the economy is populated by employed and unemployed households.

[90] Trade-union models generally find that the average tax burden may exert upward pressure on wages, but that the marginal tax burden typically reduces wages. The reason is that high marginal taxes reduce the marginal utility of increases in wages. This makes it less attractive for unions to bid for higher wages, relative to raising employment. A similar result can be derived for models of efficiency wages or search models (see e.g. Bovenberg and van der Ploeg, 1994c). Empirical evidence generally confirms this theoretical result. To illustrate, Graafland and Huizinga (1997) find an elasticity of the average tax rate of 0.6 and an elasticity of the marginal tax rate of -0.1. As the average and marginal tax rates coincide in our model, a value for γ of 0.5 seems consistent with their findings.

may exist if income during unemployment is not fully indexed to consumer wages.

7.3.4   No indexation

If income during unemployment is not indexed, we may have ß = γ = 0.5 (see subsection 7.2.2). In that case, the fourth column of Table 7.7 shows that an environmental tax reform is less favorable for employment and private income than in the previous simulation (see the third column). Indeed, the lower value-added per worker is partly borne by employers in terms of a higher producer wage. This reduces employment. As noted before, most economic models assume that $\beta = 1$. Empirically, moreover, this seems the most plausible value.

7.3.5   Wage curve

Layard et al. (1991) provide estimates for the impact of unemployment on wages for a large number of OECD countries. They find a variety of coefficients between -1 and -14, which suggest that labor markets are characterized by neither perfect competition nor strong rigidities. For the Netherlands, Graafland and Huizinga (1997) find a long-term semi-elasticity of the unemployment rate between -1.5 and -2.8. The fourth, fifth and sixth columns in Table 7.7 therefore present the same simulations as before, but now with a wage curve, i.e. $\delta = 2$.

We see that the wage curve affects our results in quantitative terms. In particular, competitive forces on the labor market imply that each expansion in employment raises the wage rate, thereby mitigating the effect on labor demand. To illustrate, the fifth column of Table 7.7 reveals that, with ß = γ = 1, employment expands by 0.37%, which is smaller than without a wage curve (compare the fourth and first columns of Table 7.7). In qualitative terms, the wage curve does not change the results of the previous experiments. Indeed, an environmental tax reform yields a triple dividend in all three cases.

A second observation from the experiments with $\delta = 2$ is that market forces make the significance of real wage resistance less pronounced. Indeed, if we take ß = 1 and γ = 0.5, the sixth column of Table 7.7 reveals that employment rises by 0.63%. This is only slightly larger than the effect on employment without real wage resistance (ß = γ = 1), in which case employment expands by 0.37%. In the model without a wage curve, real wage resistance more than triples the positive impact on employment.

A final conclusion from the experiments with δ = 2 is that profits fall in all simulations. Indeed, the increase in employment boosts wages, thereby reducing profit incomes. Accordingly, the reform shifts the incomes from the owners of the firm towards workers.

Table 7.7: Effects of an environmental tax reform that raises the producer price of energy by 10% in the model with a non-competitive labor market

	δ = 0	δ = 0			δ = 2			δ → ∞
	ß = 0	ß = 1	ß = 1	ß = ½	ß = 1	ß = 1	ß = ½	
	γ = 0	γ = 1	γ = ½	γ = ½	γ = 1	γ = ½	γ = ½	
Employment	1.3	0.57	1.98	0.99	0.37	0.63	0.47	0.33
Pollution	−5.12	−6.02	−5.33	−5.50	−5.67	−5.54	−5.57	−5.59
Consumer wage	0	0.84	−0.56	0.35	1.43	1.17	1.3	1.56
Profits	2.01	−0.77	4.18	0.85	−1.25	−0.34	−0.88	−1.35
Private income	0.4	0.14	0.64	0.29	0.08	0.17	0.11	0.06

7.3.6   Competitive labor market

The final column in Table 7.7 shows the simulation results in case of a competitive labor market, i.e. with δ → ∞. We see that wages rise by 1.56%, while profits fall by 1.35%. Hence, tax shifting from wages towards profits is important. Higher after-tax wages stimulate labor supply through the substitution effect, while lower profits encourage labor supply through the income effect. Accordingly, employment in the new equilibrium rises by 0.33%. Aggregate private incomes rise by 0.06%. Hence, an environmental tax reform yields a strong double dividend. Note that there is no pink dividend in this case, since the labor market is in equilibrium.

7.3.7   Results from other studies

The analysis in this chapter may help to better understand the simulation results from various applied studies on the double dividend. Indeed, large-scale empirical models that have been employed in the double-dividend debate yield a variety of effects. To illustrate, Table 7.8 presents the results from a number of empirical studies that calculate the effect of an energy tax reform in the EU on employment. The differences in the results may be understood partly if we know something about the elasticities in the wage equations in these models.

The simulations reported in Table 7.8 suggest that employment rises on account of an energy tax reform if the revenues are used to cut employer's social security contributions. Indeed, according to simulations with QUEST, the increase in employment ranges from 0 to 1%. The HERMES model predicts an increase in employment between 0.3% and 0.5%. One major explanation for this favorable effect on employment is that there exists real wage resistance in these models (i.e. the case with $\gamma < 1$ in our model). To illustrate, the long-term elasticity of the employer's tax in QUEST is about 0.5 (see EC, 1994, footnote 9). Accordingly, lower SSC for employers tends to reduce real wage costs. The indirect taxes on energy, in contrast, are fully borne by workers in terms of a lower real consumer wage (i.e. $\beta = 1$). In both QUEST and HERMES, a reform from income taxes to energy taxes is less favorable for employment. This is because real wage resistance is unimportant in case of income taxes ($\gamma = 1$). Hence, the long-term elasticity of the employer's tax differs from the tax burden on employees.

These findings differ from those derived with the WARM model. In particular, the wage equation in WARM is based on estimates by Brunello (1996) which suggest that real wage resistance is important only in the short run. Indeed, using WARM, Carraro et al. (1996) find that wages typically fall in the short run because of short-term real wage resistance. A carbon tax, according to the EU proposal of 1992, may thus boost employment between 0.2 and 1.3% in the EU countries. After ten years, however, wages and employment tend to revert to their baseline, since real wage resistance is unimportant in the long term.

Conrad and Schmidt (1997) adopt the GEM-E3 model in which they distinguish different labor market regimes. In particular, if consumer wages are fixed (i.e. $\beta = \gamma = \delta = 0$ in our model), they find that employment may expand by almost 1% on account of an environmental tax reform. The main explanation for this result is that the reform shifts the tax burden from labor towards other sources of income, including capital income. If the labor market would be perfectly competitive (i.e. $\delta \to \infty$ in our model), Conrad and Schmidt find that the positive effect on employment is 0.54%. This is consistent with our finding that market forces tend to reduce the magnitude of the employment effect.

Table 7.8 Effects of an EU energy tax reform on employment in different empirical models

reference	tax design	model	revenue recycling	short-run effect[a]	long-run effect[a]
European Economy (1992)	$10 per barrel of oil equivalent	HERMES	income tax		0
		HERMES	SSC[a]		0.3
		QUEST	income tax		−0.3
		QUEST	SSC[a]		0
Standeart (1992)	$10 per barrel of oil equivalent	HERMES	income tax	0	−0.1
		HERMES	SSC[a]	0.2	0.5
EC (1994)	$10 per barrel of oil equivalent	QUEST	SSC[a]	0.3	1.0
Brunello (1996)	19 ECU per ton of carbon	WARM	payroll tax	3	1

| Carraro et al. (1996) | $10 per barrel of oil equivalent | WARM | payroll tax | 0.2 to 1.3 | 0 |
| Conrad and Schmidt (1997) | $22 ECU per ton of carbon | GEM-E3 | SSC[a] | 1.00 | 0.5 |

[a] Social Security Contributions paid by employers

7.4   Conclusions

This chapter investigates the effects of an environmental tax reform in the presence of a fixed factor and an imperfect labor market. The presence of a fixed factor raises the opportunities for a double dividend, since an environmental tax reform may shift the tax burden from workers towards the owners of capital, thereby raising employment (see also chapter 4). Labor market imperfections may also raise the opportunities for an environmental tax reform to boost employment. In particular, we derive three major conclusions from our analysis. First, if income from unemployment is indexed to after-tax wages, the effect on employment depends on the production structure. More specifically, we find that employment expands if, compared to the fixed factor, labor is a better substitute for pollution. Empirical evidence indeed suggests that labor and energy are better substitutes for each other than are capital and energy. Second, if income during unemployment is indexed to value-added per worker, there exists real wage resistance. This occurs if the unemployed are able to earn untaxed incomes, e.g. income from informal activities or household production. Real wage resistance yields an additional channel through which an environmental tax reform may raise employment. In particular, by widening the gap between the utilities from employment and unemployment, lower labor taxes reduce the bargaining power of the union in wage negotiations. Accordingly, wages fall and employment expands. Empirical evidence on wage formation suggests that real wage resistance is important for the Netherlands, especially in the short run. This conclusion is similar to that in chapter 4, which also suggests that an environmental tax reform is more likely to yield a double dividend in the short run. However, whereas this result in chapter 4 depends on the degree of capital mobility, the conclusion in this chapter depends on the presence of short-term real wage resistance. Finally, this chapter shows that strong market forces tend to reduce the effect on employment in quantitative terms. However, such forces have no impact on the qualitative results on the double dividend.

8   Feedback effects of the environment on the economy

In the benchmark model, environmental quality enters utility as a weakly separable externality. With this specification, the welfare effects of an environmental tax reform can be divided into two separate and independent dividends: an environmental dividend and a non-environmental dividend. Indeed, the specification does not allow for feedback mechanisms from environmental quality on the behavior of households. Furthermore, environmental quality has no impact on the economy through its role as a public input into production.

The restrictive assumption concerning the nature of the environment in the benchmark model may be justified for some types of environmental externalities or for low levels of pollution. However, it can be unrealistic in other cases. This is shown by the literature on the valuation of an improved environmental quality. Some examples from this literature may illustrate the potential importance of environment-economy interactions. First, a number of valuation studies find that environmental degradation affects the productivity in forestry, fishing or agriculture (see e.g. Cropper and Oates, 1992, in their survey on the literature of environmental economics). Second, Ostro (1994), van Ewijk and van Wijnbergen (1995), and Zuidema and Nentjes (1997) provide evidence regarding the adverse health impact of air pollution, both in developing countries and in industrialized countries. A third category of valuation studies measure the benefits of environmental regulation by including so-called defense activities in the analysis. These activities are relative complements to pollution and can be thought of as private abatement activities by firms that suffer from pollution, or the consumption of health care by households whose condition is affected by air pollution. Indeed, such activities typically respond to improved environmental quality. A fourth example of feedback effects of environmental degradation on the economy involves those related to the greenhouse effect. Alfsen et al. (1992), Frankhauser (1995) and Tol (1995), among others, have estimated that the social damages from increasing CO_2 concentrations in the atmosphere can be substantial. These damages include productivity losses and different types of defense expenditures. Other categories of valuation studies that emphasize interactions between environmental quality and private behavior measure recreation benefits of pollution control (Braden and Kolstad, 1991) or analyze congestion-type externalities (Mayeres and Proost, 1997).

This chapter investigates what happens to our double-dividend analysis if we change the character of environmental quality in our benchmark model. In particular, we introduce two types of feedback mechanisms from environmental quality on the economy. First, we introduce environmental externalities in production. Improvements in environmental quality will thus affect not only household utility, but also the productivity of private production factors. The second interaction involves feedbacks of environmental quality on household behavior. In particular, if environmental quality enters the utility function in a non-separable manner, it may impact the optimal consumption pattern of house-holds.

In the presence of environment-economy interactions, the environmental dividend and the non-environmental dividend partly coincide. Indeed, changes in environmental quality may directly affect non-environmental welfare. This has important implications for the interpretation of the double dividend: feedback effects of environmental quality on the economy should be attributed to either the environmental dividend or the non-environmental dividend. By attributing them to the non-environmental dividend, this chapter finds that an environmental tax reform may produce a strong double dividend if production externalities are sufficiently strong. Furthermore, we find that non-separability of consumption externalities raises the opportunities for a double dividend if environmental quality is a relative substitute for leisure.

This chapter is related to other contributions in the literature on second-best environmental taxes in the presence of environment-economy interactions.

Ballard and Medema (1993) introduce production externalities in a 19-sector CGE model. Their analysis does not explicitly separate the environmental and non-environmental dividends, but focuses rather on the overall welfare effects of environmental taxes and abatement subsidies as alternative ways to improve environmental quality. Bovenberg and van der Ploeg (1994b) introduce a separable production externality in a simple general equilibrium framework and show that the optimal second-best pollution tax is equal to the Pigovian rate. This contrasts with the optimal pollution tax in case of a separable consumption externality, which is typically smaller than the Pigovian tax (see chapter 3). In his overview article on the double dividend, Bovenberg (1997) discusses the role of non-separable consumption externalities. He does not formally work out this issue, however. Schwartz and Repetto (1997) introduce a consumption externality that is a more-than-average substitute for leisure. In that case, they show that pollution taxes may stimulate labor supply through improved environmental quality. Beaumais and Ragot (1998) include a non-separable consumption externality that affects the allocation between green and non-green products, thereby referring to the literature on defense activities. Finally, Williams (1998b) incorporates externalities affecting human health. In particular, changes in human health may affect the time spent sick, thereby affecting the time constraint of households. Furthermore, reductions in human health may raise the consumption of medical care, thereby reducing the consumption of other commodities.

The rest of this chapter is organized as follows. Section 8.1 discusses the role of production externalities in the double-dividend debate. Section 8.2 analyzes the consequences of environmental taxes if environmental quality is non-separable from private commodities in the utility structure. Section 8.3 presents some numerical simulations with different characterizations of environmental quality. Section 8.4 concludes.

8.1   Production externalities

This section explores the effects of an environmental tax reform if the quality of the natural environment enters production. We abstract from polluting consumption commodities. Hence, environmental quality is determined by the use of polluting inputs into production alone, i.e. $M = m(E)$.

8.1.1   The model

Suppose that the production function looks as follows:

$$Y = F[h(E, L), M]$$ (8.1)

Hence, environmental quality (M) enters the production function in a weakly separable way from the two private inputs into production, labor (L) and the polluting input (E), where $h(.)$ is a linear-homogenous function. We assume that firms do not optimize with respect to the public environmental input (M). The production function in (8.1) shows that polluting inputs affect production in two ways. On the one hand, polluting inputs contribute to production as a rival, private factor of production (E). On the other hand, the use of polluting inputs harms the quality of the environment as a non-rival, non-excludable, public input (M). Through this latter channel, more polluting inputs adversely affect production.

By linearizing the production function in (8.1), we write output as:

$$\tilde{Y} = w_L \tilde{L} + w_E \tilde{E} + Y_M w_M \tilde{M}$$ (8.2)

where $Y_M \equiv \partial Y / \partial M$ denotes the marginal productive capacity of the environment and $w_M = M/Y$. Assuming constant returns to scale with respect to the private production factors, i.e. E and L, we obtain the following non-profit condition:

$$w_L(\tilde{W} + \tilde{T}_L) + w_E \tilde{T}_E - Y_M w_M \tilde{M} = 0$$ (8.3)

We assume that the rest of the model does not differ from the benchmark model. Hence, apart from output and the non-profit condition, the economy is characterized by the expressions in Table 3.1.

Following the same procedure as in appendix 3B, we derive the following expression for the *MEB*:

$$MEB = -\theta_L w_L \tilde{L} \ - \ [\,\theta_E - (\frac{u_M}{\lambda} + Y_M)\,\gamma_E\,]\,w_E\,\tilde{E}$$

(8.4)

$$\underbrace{\phantom{-\theta_L w_L \tilde{L}}}_{\substack{labor-market \\ distortion}} \qquad \underbrace{\phantom{(\frac{u_M}{\lambda} + Y_M)\,\gamma_E}}_{\substack{environmental \\ distortion}}$$

where $\gamma_E \equiv -m_E/(P_E+T_E)$ measures the marginal impact of polluting inputs on environmental quality. Comparing (8.4) with its counterpart from the benchmark model in expression (3.10), we find that the first-best tax on polluting inputs (i.e. if the labor market is not distorted so that $\theta_L = 0$) is higher in (8.4) because the Pigovian tax should internalize not only the consumption externality $(u_M\gamma_E/\lambda)$ but also the production externality $(Y_M\gamma_E)$. As in chapter 3, expression (8.4) can be used to investigate whether the optimal pollution tax will exceed the Pigovian rate in a second-best framework in which the labor market is distorted, i.e. where $\theta_L > 0$. In particular, if we start from the Pigovian tax (where $\theta_E = Y_M\gamma_E + u_M\gamma_E/\lambda$), a marginal increase in the pollution tax will not affect welfare through the channel of the environmental distortion. However, according to (8.4), it increases welfare if it raises employment. In that case, the optimal second-best pollution tax is higher than the Pigovian tax. In the benchmark model, we find that employment declines if we start from the Pigovian rate, so that the optimal second-best pollution tax is smaller than the first-best level. In this chapter we investigate whether this conclusion holds true if we include production externalities in the model.

The *MEB* in (8.4) can be written alternatively in terms of two dividends:

$$MEB = -\tilde{B} - \frac{u_M}{\lambda} w_M \tilde{M}$$

$$\begin{array}{cc} blue & green \\ dividend & dividend \end{array}$$

(8.5)

where

$$\tilde{B} \equiv \theta_L w_L \tilde{L} + \theta_E w_E \tilde{E} + Y_M w_M \tilde{M} \tag{8.6}$$

As in chapter 3, the marginal excess burden in expression (8.5) is decomposed into two separate dividends, namely a green dividend and a blue dividend. In contrast to chapter 3, the choice for this decomposition is not directly obvious. Indeed, a cleaner environment improves welfare through two different channels. First, it yields environmental amenities, as environmental quality enters the utility function. Second, it reduces the adverse production externalities because environmental quality enters the production function. In expression (8.5), we attribute the environmental amenities to the green dividend. In contrast, the welfare improvements due to lower adverse production externalities are attributed to the non-environmental dividend. In particular, expression (8.6) suggests that the blue dividend is determined by, on the one hand, the effects on the respective tax bases of labor and pollution taxes and, on the other hand, the productivity gain associated with a cleaner environment. The reason why we include the welfare effect associated with negative production externalities in the blue dividend is that these productivity gains accrue to private households in terms of higher private incomes. This contrasts with the environmental amenities, which are public and thus leave private incomes unchanged.

The blue dividend in (8.6) can be written alternatively by using the government budget constraint in (T3.8) and the non-profit condition in (8.3):

$$\tilde{B} \equiv (1 - \theta_L) w_L \tilde{W} \tag{8.7}$$

Expression (8.7) reveals that the benefits of a cleaner environment (through their role as a public input into production) are reflected in a higher after-tax wage rate.[91]

By solving the model along the lines of appendix 3C, and using (8.2) and (8.3) instead of (T3.1) and (T3.3), we derive the reduced-form equations for employment, private incomes and pollution in the presence of production externalities. These are presented in Table 8.1.

8.1.2   Environmental tax reform

If we start from an equilibrium without pollution taxes (i.e. $\theta_E = 0$), the reduced forms in the first and third rows of Table 8.1 show that an environmental tax reform boosts employment and private incomes if $Y_M \gamma_E > 0$. In particular, by reducing the demand for polluting inputs, environmental taxes improve the quality of the natural environment if $\gamma_E > 0$. Because the environment acts as a non-rival, non-excludable public input, private production factors become more productive if $Y_M > 0$. Since the price of polluting inputs is fixed on the world market, this productivity gain accrues to workers in the form of a higher labor productivity. This raises the wage rate and stimulates the incentives to supply labor. Accordingly, employment rises. We call this effect of an environmental tax reform the *environmental productivity effect*.

If we raise pollution taxes in an equilibrium where these taxes are initially positive (i.e. $\theta_E > 0$), the positive productivity effect should be weighed against the tax burden effect (discussed in section 3.3). In particular, an environmental tax reform succeeds in producing a strong double dividend if a lower level of aggregate pollution substantially improves environmental quality (i.e. γ_E is large) and a cleaner environment exerts a large impact on productivity (i.e. Y_M is large). At the same time, the initial pollution tax (θ_E) should be small so that a reduction in pollution is cheap.

---

[91] Hence, if the production externality would be attributed to the green dividend, the blue dividend could no longer be measured by the change in after-tax wages.

From the reduced-form for employment, we find that employment expands only if the initial pollution tax is smaller than the adverse production externality of pollution, i.e. if $\theta_E < \gamma_E Y_M$. Hence, if we start from the first-best pollution tax where $\theta_E = \gamma_E Y_M + u_M \gamma_E / \lambda$, a marginal increase in the pollution tax reduces employment. Accordingly, the optimal second-best pollution tax is smaller than the Pigovian rate (see (8.4)). The production externality thus does not overturn the corresponding conclusion from the benchmark model.[92]

Table 8.1: Reduced forms in the model with separable production externalities

	$\tilde{T}_E$
$\Delta \tilde{L}$	$-\eta_{LL}[\theta_E - Y_M\gamma_E]w_E\dfrac{\sigma_E}{\Omega}$
$\Delta \tilde{E}$	$-[(1-\theta_L)w_L - \theta_L w_L \eta_{LL}]\dfrac{\sigma_E}{\Omega}$
$\Delta \tilde{B}$	$-[\theta_E - Y_M\gamma_E]w_E\dfrac{\sigma_E}{\Omega}$

$$\Delta = (1-\theta_L)w_L - \eta_{LL}[\theta_L w_L + (\theta_E - Y_M\gamma_E)w_E\frac{w_L(1+Y_M w_M)}{\Omega}] > 0$$

$$\Omega = (1+Y_M w_M)w_L + Y_M\gamma_E w_E\sigma_E$$

---

[92] The production function in (8.1) is restrictive in the modeling of the production externality. Indeed, with a separable externality, changes in environmental quality do not affect the input mix between labor and polluting inputs. This is unrealistic if, for instance, pollution affects human health, but not the productivity of other inputs. If, compared to the polluting input, labor is a relative complement to environmental quality, a higher supply of environmental quality would make labor-intensive production more attractive. In our model, however, the assumption of non-separable production externalities would imply that profits emerge. Accordingly, the benefits of less production externalities accrue to profits and wages.

8.1.3   Optimal environmental tax

By substituting the reduced forms for employment and pollution from Table 8.1 into (8.4), we find the reduced form for the *MEB*. By setting this to zero, we derive for the optimal pollution tax:

$$\theta_E = \frac{1}{\eta} \frac{u_M \gamma_E}{\lambda} + Y_M \gamma_E \qquad (8.8)$$

$$\eta = \frac{1}{1 - T_L \eta_{LL}} \qquad (8.9)$$

Expression (8.8) reveals that there is an important difference between using pollution taxes to internalize the production externality (the first term on the RHS) and the consumption externality (the second term on the RHS). Indeed, contrary to the production externality, the consumption externality in (8.8) is divided by the marginal cost of public funds (MCPF), defined in (8.9). Hence, preexisting labor-market distortions, as reflected by a higher MCPF, reduce the optimal pollution tax as an instrument to internalize the consumption externality, but do not render the pollution tax less attractive as an instrument to internalize the production externality (see also Bovenberg and van der Ploeg, 1994b).[93] Intuitively, in contrast to consumption externalities, the benefits of fewer production externalities accrue to private agents in the form of higher wage incomes. Indeed, fewer negative production externalities raise productivity, thereby increasing wages as well. This stimulates the incentives for labor supply. The benefits of fewer consumption externalities, however, are public: they

---

[93] This result is reminiscent of the production efficiency result of Diamond and Mirrlees (1971). Williams (1998b) explores the optimal tax on a good that causes production externalities in a somewhat different framework. In particular, his model includes a fixed factor in production and assumes that profit taxes are not available. In that case, he finds that the optimal second-best pollution tax is typically smaller than the Pigovian rate. This is because the productivity gain associated with a cleaner environment partly accrues to the owners of the fixed factor, rather than to workers. Accordingly, the positive effect of a cleaner environment on labor supply is mitigated. De Mooij et al. (1998) find a similar result for public expenditures on infrastructure.

benefit all, irrespective of labor supplied. Hence, they leave labor supply and private incomes unchanged.

8.2   Non-separable consumption externalities

The model in chapter 3 assumes that the quality of the natural environment enters the utility function in weakly separable form. This implies that changes in environmental quality neither affect the consumption/leisure choice nor the distribution of consumption over polluting and clean commodities. This assumption is rather restrictive. For some types of externalities, such as congestion or those affecting recreation demand, households typically do respond to changes in the environment. Hence, it is important to investigate how important the separability assumption is for the double-dividend analyses. Indeed, one might argue that -- especially in the long run -- the environment in which we live determines to a large extent our choices, our habits and the activities we undertake.

This section analyzes environmental tax reforms when the utility structure is non-separable in environmental quality. To be able to derive some analytical results, we consider a particular structure of the utility function. Furthermore, this section abstracts from polluting inputs into production or from environmental production externalities, i.e. $M = m(D)$.

8.2.1   The model

Suppose the utility structure looks as follows:

$$U = u[G, V, h(M, q[C, D])]$$ (8.10)

Hence, clean and polluting consumption commodities are first combined to yield a composite consumption commodity (Q). Then, this composite combines with environmental quality to yield sub-utility H. Finally, combining H with a composite of leisure and public consumption produces utility.[94]

From utility maximization, we can derive the first-order conditions (3.5) - (3.7). Linearizing these, we find the following expression for labor supply:

$$\tilde{L}_S = \eta_{LL}\,\tilde{W}_R + \psi(\frac{\sigma_V}{\sigma_M} - 1)\frac{u_M w_M}{\lambda}\tilde{M} \tag{8.11}$$

$$\textit{where}\quad \eta_{LL} = \frac{\sigma_V - 1 - \dfrac{u_M w_M}{w_H}(\dfrac{\sigma_V}{\sigma_M} - 1)}{\dfrac{1}{V} + \dfrac{u_M w_M}{w_H}(\dfrac{\sigma_V}{\sigma_M} - 1)},\qquad \psi = \frac{\dfrac{1}{w_H}}{\dfrac{1}{V} + \dfrac{u_M w_M}{w_H}(\dfrac{\sigma_V}{\sigma_M} - 1)}$$

and $w_H = w_C + w_D + u_M w_M$. The parameter σ_V stands for the substitution elasticity between leisure (V) and the composite (H), and σ_M is the substitution elasticity between environmental quality (M) and the consumption bundle (Q). The first term on the RHS of (8.11) shows the impact of a higher real after-tax wage rate (W_R) on labor supply. We assume that the uncompensated wage elasticity (η_{LL}) is positive so that higher real wages boost labor supply. The second term on the RHS of (8.11) shows the direct impact of a cleaner environment on labor supply. In particular, if environmental quality is an equally good substitute for leisure as it is for consumption (i.e. $\sigma_V = \sigma_M$), then it exerts no effect on labor supply. Indeed, in that case changes in environmental quality do not impact the marginal rate of substitution between consumption and leisure. This case is equivalent to the separable consumption externality discussed in chapter 3. If, compared to consumption, leisure is a relative complement to environmental quality (i.e. $\sigma_M > \sigma_V$), a cleaner environment raises the marginal utility of leisure more than the marginal utility of consumption. Accordingly, the optimal consumption/leisure ratio drops. Improvements in environmental quality

---

[94] The specification in (8.10) is attractive because it allows us to derive intuitive analytical results in the presence of a non-separable consumption externality. Other CES structures might be equally valid, but expressions become less intuitive.

thus harm the incentives to supply labor. In contrast, if leisure is a better substitute for environmental quality than consumption is (i.e. $\sigma_V > \sigma_M$), a cleaner environment makes consumption more enjoyable compared to leisure. Hence, pollution taxes crowd out leisure more substantially than consumption does, so that labor supply expands.[95]

Table 8.2: Reduced forms in the model with non-separable consumption externalities

	$\tilde{T}_D$
$\Delta \tilde{L}$	$-[\,\eta_{LL}\theta_D w_D - (1-\theta_L)w_L\,(\dfrac{\sigma_V}{\sigma_M} - 1)\,\dfrac{u_M\gamma_D}{\lambda}\,\psi w_D\,]\,\dfrac{w_C}{w_C + w_D}\,\sigma_{CD}$
$\Delta \tilde{D}$	$-[\,(1-\theta_L)w_L\ -\ \eta_{LL}\theta_L w_L\,]\,\dfrac{w_C}{w_C + w_D}\,\sigma_{CD}$
$\Delta \tilde{B}$	$-(1-\theta_L)w_L\,[\,\theta_D w_D - \theta_L w_L\,(\dfrac{\sigma_V}{\sigma_M} - 1)\,\dfrac{u_M\gamma_D}{\lambda}\,\psi w_D\,]\,\dfrac{w_C}{w_C + w_D}\,\sigma_{CD}$

$$\Delta = [(1-\theta_L)w_L - \theta_D w_D] - \eta_{LL}[\theta_L w_L + \theta_D w_D] + w_L(\frac{\sigma_V}{\sigma_M} - 1)\frac{u_M\gamma_D}{\lambda}\psi w_D > 0$$

---

[95] Parry (1995), Håkonsen (1996) and Orosel and Schöb (1996) argue that an environmental tax reform may affect the opportunities for the double dividend also if leisure is non-separable from the two private consumption commodities. In particular, referring to the Corlett-Hague rule of optimal taxation, they show that an environmental tax reform may stimulate labor supply if, compared to clean commodities, polluting commodities are poorer substitutes for leisure (Corlett and Hague, 1953).

8.2.2   Environmental tax reform

The expression for the *MEB* does not change compared to chapter 3; i.e., expressions (3.10) and (3.11) continue to represent the welfare effects of the environmental tax reform. By solving the model along the lines of appendix 3C - - where we use (8.11) instead of (T3.6) -- we derive the reduced forms for employment, pollution and blue welfare. These are presented in Table 8.2. If we start from an equilibrium without environmental taxes (i.e. $\theta_D = 0$), Table 8.2 reveals that employment and private incomes may either rise or fall, depending on the substitution elasticities in utility. In particular, if leisure and environmental quality are relative substitutes (i.e. $\sigma_V > \sigma_M$), then an environmental tax reform raises employment and private incomes because a cleaner environment encourages households to supply more labor. Hence, by enhancing both green welfare and blue welfare, an environmental tax reform yields a strong double dividend (see (8.4)). In contrast, if leisure and environmental quality are relative complements, then environmental taxes reduce the incentives to supply labor by raising the supply of the public environmental good. Accordingly, employment and private incomes fall and the strong double dividend fails. Hence, compared to the separable externality in the benchmark model, the non-separable character of the consumption externality provides both dangers and opportunities for the double dividend.[96]

If we start from an equilibrium with positive pollution taxes (i.e. $\theta_D > 0$), Table 8.2 shows that an environmental tax reform exerts two effects on employment and private incomes. Indeed, apart from the effect on the consumption/leisure choice, environmental taxes exert a tax burden effect, as discussed in section 3.2. A double dividend is feasible only if environmental quality encourages labor supply, i.e. if both leisure is a better substitute for

---

[96] Another interesting example would be a non-separable consumption externality that affects the allocation between clean and dirty commodities. Activities that are likely to fall on account of a cleaner environment are defensive expenditures such as cleaning services, health care or private abatement activities.

environmental quality than consumption, and the tax burden effect is small. This is possible for small levels of the pollution tax.

8.2.3   Optimal environmental tax

Substituting the reduced forms for private income and polluting consumption from Table 8.2 into the *MEB* in (8.5), we derive the optimal pollution tax by setting the *MEB* equal to zero:

$$\theta_D = \frac{1}{\eta} \frac{u_M \gamma_D}{\lambda} \tag{8.12}$$

$$\eta \equiv \frac{1}{1 - \eta_{LL} T_L + \theta_L w_L \psi \left(\dfrac{\sigma_V}{\sigma_M} - 1 \right)} \tag{8.13}$$

The optimal tax in (8.12) is similar to the benchmark model. Indeed, it amounts to the Pigovian rate, divided by the MCPF defined in (8.13). Without interactions between environmental quality and household behavior (i.e. $\sigma_V = \sigma_M$), the MCPF exceeds unity if the labor tax is positive (i.e. $T_L > 0$) and the uncompensated labor-supply elasticity is positive (i.e. $\eta_{LL} > 0$). If environmental quality affects the consumption/leisure choice, however, the MCPF may be smaller than unity. Indeed, expression (8.13) reveals that this may happen if, compared to consumption, leisure is a much better substitute for environmental quality (i.e. $\sigma_V > \sigma_M$). In that case, the optimal pollution tax is higher than the Pigovian rate. This contrasts with the conclusion from the benchmark model, which suggests that the optimal tax is smaller than the Pigovian rate. Intuitively, if we start from the Pigovian rate, a further increase in the pollution tax may render the tax system more, rather than less, efficient. This is because the boost in labor supply on account of a better environmental quality dominates the reduction in labor supply associated with lower real after-tax wage incomes.

This result is closely related to the literature on the welfare effects of public goods in the presence of distortionary taxes. In particular, Wildasin (1984) and Topham (1984) explore how the MCPF varies with the way in which a particular public good enters utility. If the public good is separable from private consumption and leisure, they find that the MCPF exceeds unity if the uncompensated labor-supply elasticity is positive. This result is similar to (8.13) if environmental quality is an equally good substitute for leisure and private consumption. If public goods are closer substitutes for private consumption than they are for leisure, the MCPF is even larger. In that case, raising the supply of public goods is even more expensive than if public goods are separable from private goods. However, public goods that are relatively complementary to private consumption may reduce the MCPF, possibly below unity.

8.3 Numerical simulations

This section illustrates the theoretical findings from the previous two sections numerically. In calibrating the model, we adopt the same parameters as in the benchmark model (see Table 3.3) and run simulations for two versions of the model. First of all, we analyze the model with a separable production externality. As empirical information on production externalities is scarce or applies to specific environmental externalities, we explore different levels of the marginal product of environmental quality. These simulations are presented in Table 8.3. Next, Table 8.4 shows the simulations of the model with non-separable consumption externalities.

8.3.1 Separable production externalities

Table 8.3 shows the effects of an energy tax that raises the energy price of firms by 10% in the model with separable production externalities. The parameter γ_E is set to unity, while the value of Y_M is varied. The table reveals that production externalities reduce the non-environmental cost of pollution taxes as compared to the benchmark model (which are represented by the first column of Table 8.3). In particular, the second column in Table 8.3 shows that, in the presence of small production externalities (i.e. $Y_M = 0.01$), employment and private incomes fall less than in the benchmark model (see the first column of Table 8.3 where Y_M

= 0). Note that this boost in productivity renders the fall in pollution also smaller. Large production externalities may even imply a double dividend. Indeed, the third column of Table 8.3 shows that, for $Y_M = 0.1$, an environmental tax reform boosts employment by 0.13% and private income by 0.26%. At the same time, pollution falls by 4.72%.

Table 8.3: Effects of an energy tax that raises energy prices for firms by 10% in the model with separable production externalities (with $\gamma_E = 1$ and different levels of Y_M)

	$Y_M = 0$	$Y_M = 0.01$	$Y_M = 0.1$
Employment	-0.07	-0.04	0.13
Pollution	-5.07	-5.02	-4.72
Private income	-0.14	-0.09	0.26

Table 8.4: Effects of an energy tax that raises energy prices for households by 10% in the model with non-separable consumption externalities (with $\sigma_V = 1.5$ and different levels of σ_M)

	$\sigma_M = 0.5$	$\sigma_M = 1.5$	$\sigma_M = 4.5$
Employment	0.01	-0.05	-0.07
Pollution	-3.90	-4.06	-4.12
Private income	-0.07	-0.10	-0.12

8.3.2   Non-separable consumption externalities

Table 8.4 shows the effects of an energy tax that raises energy prices of households by 10% in the model with non-separable consumption externalities. We keep σ_V at 1.5, which is similar to the value of the benchmark model. This value differs from σ_M. Table 8.4 reveals that, if environmental quality and consumption are poor substitutes (i.e. $\sigma_M = 0.5$), an environmental tax reform may stimulate employment. By expanding the labor-tax base, this mitigates the negative effect of the reform on private income. If, compared to leisure, consumption is a better substitute for environmental quality (i.e. if $\sigma_M = 4.5$), then pollution taxes reduce employment. By further eroding the tax base, this reduction in employment renders the decline in private income even larger than in the case of separable consumption externalities.

8.4   Conclusions

This chapter shows that, in the presence of production externalities or if consumption externalities are relatively good substitutes for leisure, an environmental tax reform may stimulate labor supply. Accordingly, the double dividend may be feasible. However, if consumption externalities are complementary to leisure, then a cleaner environment reduces labor supply even more than with separable externalities. Hence, interactions between environmental quality and private behavior provide both opportunities and dangers for a double dividend.

Most economic analyses -- including our benchmark model -- ignore feedback mechanisms from the environment on the economy. The reason is that clear-cut empirical evidence on these feedbacks is scarce for broad measures of environmental quality. Analyzing interactions, however, is important for at least two reasons. First, investigating feedback mechanisms can teach us how robust the outcomes from economic analyses are with respect to particular separability assumptions. Second, empirical evidence on specific externalities suggests that ignoring interactions is often too restrictive. Indeed, a number of valuation studies suggests that improved environmental quality may affect human health, production methods or consumption patterns. In order to investigate the opportunities for a double dividend for those externalities, we need to incorporate interactions between the environment and the economy.

The consequences of feedback mechanisms for the double-dividend analysis should be interpreted with caution. In particular, in a second-best framework, curbing pollution may become more favorable for welfare than would be the case without these feedback effects. This is because environmental improvements induce favorable behavioral responses that alleviate preexisting tax distortions. However, this result holds for all environmental policy instruments. Indeed, it is the environmental dividend itself that is responsible for the existence of the other, non-environmental dividend. The non-environmental dividend, thus, does not originate in the recycling of the revenues of the environmental tax, which is usually put forward as the main argument in favor of the double-dividend claim. This result implies that environmental quality is the key towards a successful environmental tax reform, including a double dividend.

9   Green tax reform in an endogenous growth model

The previous chapters have explored the double dividend in a static framework. This chapter extends the analysis by employing a dynamic model of endogenous growth. The model builds on an endogenous growth model developed by Barro (1990) and is used to analytically explore the link between environmental externalities and distortionary income taxes. Other authors have employed numerical analyses of the double dividend in an intertemporal framework with exogenous growth (see e.g. Goulder, 1995b and Proost and van Regemorter, 1995).

This chapter is related to the literature on pollution and long-term growth (see e.g. Bovenberg and Smulders (1995), Den Butter and Hofkes (1993), Gradus and Smulders (1993), and Ligthart and van der Ploeg (1995)). This literature focuses on optimal environmental policies and explores how the social optimum can be sustained in a decentralized economy. Our analysis departs from this first-best world in two major ways. First, we explore the consequences of a reform of the tax system, starting from an initial equilibrium that is not necessarily optimal. Second, in addition to market failures associated with environmental externalities, we allow for tax distortions due to the absence of lump-sum taxation.

Whether an environmental tax reform produces a second, non-environmental dividend depends on how such a reform impacts the growth rate. In particular, by driving a wedge between the marginal social costs and benefits of capital accumulation, distortionary income taxes reduce economic growth below its first-best level. Accordingly, a reform that boosts growth alleviates the deadweight loss from the distortionary income tax. In this way, an environmental tax reform may yield a double dividend, i.e. not only an increase in welfare from environmental amenities, but also an increase in welfare from private commodities.

We find that an environmental tax reform harms growth, and hence private welfare, if two conditions are met. First, the positive externality of a better environmental quality on productivity should be small compared to the production elasticity of pollution as a rival input into production. Second, substitution between pollution and other inputs should be rather easy. If these two conditions are met, the government faces a trade-off between environmental care and economic growth. If substitution between pollution and other inputs is difficult, then pollution taxes are less powerful in cutting pollution. At the same time, they are a more effective device to tax the quasi-rents from pollution, thereby generating revenues to reduce the distortionary tax on output. Accordingly, an environmental tax reform may raise growth and thus private welfare by enhancing the efficiency of the tax system as a revenue-raising device. We show that the optimal environmental tax may exceed the Pigovian level if substitution is difficult.

The rest of this chapter is organized as follows. Section 9.1 discusses the model. Section 9.2 explores the effects of pollution taxes on growth, environmental quality and welfare if the substitution elasticity between capital and pollution equals unity. The case with small substitution possibilities between capital and pollution is investigated in section 9.3. Section 9.4 discusses the effects of environmental policy instruments that do not charge a price for pollution. Finally, section 9.5 concludes.

9.1   The model

In endogenous growth models, man-made commodities grow at endogenous growth rates while other inputs are constant. In our model, growth in output is sustainable because it is consistent with a fixed level of environmental quality.[97]

---

[97] In most endogenous growth models, the inputs that are not man-made take the form of raw labor. In our model, however, these inputs are supplied by nature. For another endogenous

Indeed, after a policy shock, the economy immediately moves towards a so-called balanced growth path on which all man-made goods grow at the same rate while natural variables, including the quality of the environment, remain at a constant level. Transitional dynamics in the ratios between the man-made variables are absent because the model includes only one type of asset, namely capital.[98]

Balanced, sustainable growth should be both feasible and optimal. For growth to be feasible, the production function must meet the so-called 'core' property (see Rebelo (1991)), according to which capital is produced with a constant-returns technology in man-made inputs. Moreover, for the natural inputs to play a non-trivial role in the long run, various substitution elasticities must be unity. The requirement that balanced growth with a constant level of environmental quality is not only feasible but also optimal imposes restrictions on the utility function (see below). The rest of this section presents the model in more detail.

9.1.1   Structure of the model

Production function
We extend the production technology in Barro (1990) and Barro and Sala-i-Martin (1992) by including pollution, abatement, and environmental quality as inputs into production. In particular, production technology is described by:

---

growth model in which growth of output is sustainable (i.e. consistent with a constant level of environmental quality), see Gradus and Smulders (1993).

[98] Alternatively, the quality of the environment could be modeled as a stock rather than as a flow (see e.g. Bovenberg and Smulders (1995)). However, this would complicate the model substantially, as it would imply a second state variable, thereby introducing transitional dynamics. More importantly, the qualitative steady-state results do not change if a stock, rather than a flow, determines the external effects (see Smulders and Gradus (1996), appendix A). Hence, if one is interested in long-term effects, the assumption of a flow of pollution involves no loss of generality. In more complicated models with stock-flow dynamics, issues of stability and existence of a steady state may arise (Tahvonen and Kuuluvainen (1993)). However, stability and existence can be established in simple models that include environmental quality as a stock (see Bovenberg and Smulders (1996)).

$$Y = f[h(K, S), n(A, P)] \, M^{\eta}$$

$$\text{where} \quad n(A, P) = A P^{\alpha}$$

(9.1)

where η denotes the environmental externality parameter in production. According to (9.1), firms combine three private inputs with two public inputs to produce output. On the one hand, following Barro (1990) and Barro and Sala-i-Martin (1992), physical capital (K) and productive government spending (S), which can be thought of as infrastructure, are combined to produce an intermediate input, H. The function $h(K,S)$ exhibits constant returns to scale with respect to the two inputs. On the other hand, by combining pollution (P)[99] and private abatement (A), firms produce another intermediate input, N.[100] The function $n(A,P)$ features constant returns with respect to the man-made input of abatement, which, in contrast to the natural input of pollution, is growing on a balanced-growth path. Together with the public services from the environment (M)[101], the two intermediate inputs, H and N, yield output (Y), where the production function $f(H,N)$ exhibits constant returns with respect to H and N.

For balanced growth to be feasible, the production function must meet a number of restrictions.[102] First, the production function must meet the core property by exhibiting constant returns with respect to the growing inputs:

---

[99] Pollution, P, differs from the polluting input discussed in previous chapters in that its market price is zero. In the present framework, pollution can alternatively be modeled as an output (see, e.g., van der Ploeg and Withagen (1991), Gradus and Smulders (1993) and Smulders and Gradus (1996)). Siebert, Eichenberger, Gronych and Pethig (1980) show that these two modelling approaches are equivalent. In our model, pollution can be interpreted as a polluting input with zero extraction costs. Emissions are proportional to this input.

[100] Abatement is assumed to be a private input that enables firms to increase output without causing more pollution. Other studies take abatement as a public, rather than a private, input (see e.g. Nielsen, Pedersen and Sørensen, 1995; Bovenberg and van der Ploeg, 1994ab; van der Ploeg and Bovenberg, 1994; Ligthart and van der Ploeg, 1998). In that case, abatement can be viewed as knowledge about 'clean' production methods.

[101] Pollution can be viewed as the extractive, rival use of the environment. At the same time, the environment, M, yields non-extractive, non-rival services as an input into production.

[102] Smulders and Gradus (1996, sections 3 and 4) and Bovenberg and Smulders (1995, section 3) formally derive the necessary conditions for balanced growth.

$$\delta + \beta + \gamma = 1 \qquad\qquad (9.2)$$

where δ, β and γ denote the production elasticities of the growing inputs S, K and A, respectively.

The second condition for balanced-growth is that the production elasticities of the various inputs on the LHS of (9.2), as well as the production elasticity of pollution ($\alpha\gamma$), remain constant over time. This imposes restrictions on the substitution possibilities between the various inputs. In particular, on a sustainable growth path, the flow of pollution (P) is constant, while abatement (A) grows at the same rate as output. In the face of these diverging growth rates, the production elasticities of pollution and abatement (i.e. $\alpha\gamma$ and γ) remain constant only if the elasticity of substitution between the non-growing input, P, and the growing input, A, equals unity. In particular, if substitution would be too easy (i.e. the substitution elasticity > 1), the production elasticity of pollution would approach zero, as the growing input (abatement) would crowd out the non-growing input (pollution). If substitution would be too difficult (i.e. the substitution elasticity < 1), in contrast, the production elasticity of abatement would approach zero, because the marginal product of abatement would fall substantially as the abatement-pollution ratio rises. For the same reasons, the elasticity of substitution between $f(N,H)$, which grows on the balanced-growth path, and the quality of the environment, which remains constant, should equal unity. Finally, the substitution elasticity between K and S, denoted by σ_H, as well as the substitution elasticity between H and N, denoted by σ_Y, should be constant. However, these elasticities are allowed to differ from unity as the arguments in h and f (i.e. K, S and H, N, respectively) grow at the same constant rate.[103]

---

[103] If abatement would not enter the model, the substitution elasticity between (constant) pollution and the (growing) intermediate input, H, would be restricted to unity (see the analysis in section 9.2). Incorporating abatement thus enables us to examine alternative substitution possibilities. Section 9.3 indicates that this is potentially important.

Firm behavior

The model describes a decentralized economy with perfect competition. A representative firm maximizes the firm value with respect to abatement, pollution and investment in physical capital. The firm ignores the environmental production externality. The value of the firm amounts to the present value of all future dividends, i.e.:

$$V = \int_0^\infty D e^{-Rt} dt \tag{9.3}$$

Dividends are represented by:

$$D = (1 - T_Y) Y - (1 - T_A) A - T_P P - I \tag{9.4}$$

where T_Y stands for a tax on output, T_A represents an abatement subsidy and T_P denotes a pollution tax. Units are normalized such that the prices for output and abatement are unity. The interest rate, denoted by R, is constant on a balanced-growth path. We abstract from depreciation of capital. Hence, the accumulation of capital ($\dot{K} = dK/dt$) equals gross investment (I), i.e.:

$$I \equiv \dot{K} \tag{9.5}$$

Profits (Π) are defined as dividends plus investment minus capital costs, i.e.:[104]

$$\Pi = D + I - RK \tag{9.6}$$

Positive profits can exist in equilibrium due to the presence of a fixed factor (e.g., land, know-how, managerial talent), which constitutes a barrier for entering

---

[104] Non-negative firm profits (i.e. $w_{\Pi} = (1 - T_Y)(\delta - \alpha\gamma) \geq 0$) require that the production elasticity of pollution at the firm level ($\alpha\gamma$) does not exceed the production elasticity of public investment (δ). With production featuring constant returns with respect to the growing inputs, S, A and K (see (9.2)), the non-negativity condition for profits guarantees that the production function does not exhibit increasing returns with respect to the 'private' inputs P, A and K. Merger across firms is thus not beneficial and a competitive equilibrium can be sustained. Hence, perfect competition is possible only if public infrastructure enters the production function (so that $\delta > 0$).

firms. This chapter does not incorporate a direct tax on these profits. Solving the maximization problem of the firm, we find the implicit demand equations for private inputs:

$$(1 - T_Y)\frac{\partial Y}{\partial A} = (1 - T_A) \tag{9.7}$$

$$(1 - T_Y)\frac{\partial Y}{\partial P} = T_P \tag{9.8}$$

$$(1 - T_Y)\frac{\partial Y}{\partial K} = R \tag{9.9}$$

Expressions (9.7) - (9.9) reveal that firms equalize the marginal productivity of private inputs to their respective producer prices.

Household behavior
A representative household maximizes the following intertemporal utility function subject to a dynamic budget constraint:

$$Max \quad U = \int_0^\infty \frac{[CM^\phi]^{1-\frac{1}{\sigma}}}{1 - \frac{1}{\sigma}} e^{-\theta t} dt \tag{9.10}$$

subject to

$$\dot{K} = RK + \Pi - C \tag{9.11}$$

where ϕ measures the environmental consumption externality, σ stands for the intertemporal substitution elasticity and θ is the pure rate of time preference. According to (9.10), two arguments enter instantaneous utility: private consumption (*C*) and the quality of the environment (*M*). The substitution elasticity between the growing variable, *C,* and the quality of the environment, *M*, must be unity to ensure that balanced sustainable growth is optimal from a social point of view. Intuitively, on a balanced growth path, optimal saving rates and the optimal demand for environmental services should not respond to output becoming more abundant relative to environmental quality. This requires that the elasticity of marginal utility $U_{CC}C/U_C$ be constant. Moreover, the positive income effect on the demand for environmental services, which is triggered by output growth, exactly offsets the negative substitution effect on account of a rising shadow price of environmental quality. If the substitution elasticity were smaller than unity, the income effect would dominate the substitution effect so that saving rates, and hence growth, would decline as produced goods become more abundant relative to environmental quality. If the substitution elasticity between C and M would exceed unity, in contrast, saving rates and hence growth rates would increase over time, as the substitution effect would dominate the income effect. In that case, produced goods would substitute for environmental amenities in the long run so that these amenities would not be relevant for long-run utility (see Smulders and Gradus (1996) and Bovenberg and Smulders (1995)).

Maximizing household utility subject to the household budget constraint, one finds the Keynes-Ramsey rule:

$$g = \frac{\dot{C}}{C} = \sigma(R - \theta) \tag{9.12}$$

where g represents the growth rate of the economy that applies to all growing variables.[105]

Government

The government budget is balanced according to:

$$T_Y Y + T_P P = T_A A + S \tag{9.13}$$

---

[105] For the steady state to be meaningful, the consumption share of output should be non-negative. This condition ensures also that utility is bounded. From (9.5), (9.6) and (9.11), we know that $w_C = w_D$. On a balanced-growth path, D is growing at a constant rate, g. Solving the integral (9.3), we derive: $w_C = w_D = (R-g)w_V$. Hence, the growth rate, g, should not exceed the interest rate. This restricts the rate of time preference (θ) and the intertemporal elasticity of substitution (σ). In particular, from (9.12) we find that a non-negative consumption share requires that $\theta\sigma + (1-\sigma)R > 0$ (i.e., the intertemporal substitution possibilities cannot be too large, especially if the rate of time preference is small).

The government does not issue public debt and raises revenues by adopting a positive tax rate on output (T_Y), which is constant on a balanced-growth path, and a positive tax rate on pollution (T_P), which grows at the same rate as output so as to keep constant the share of pollution tax revenues in output, i.e. $T_P{}^* = T_P P/Y$. The revenues from these taxes are used to finance two types of government spending: first, public investment (S), which grows at the rate of output and, second, abatement subsidies (T_A) which are constant. The government balances its budget at each point in time by adjusting the tax rate on output.

Environmental quality
The inverse relationship between the quality of the environment and the flow of pollution is formalized by:

$$M = m(P) \qquad (9.14)$$

Environmental quality in the initial equilibrium is assumed to be above a certain ecological threshold level. It remains above this critical load, as pollution does not grow on the balanced-growth path.

Walras law
The model describes a closed economy. The equilibrium condition on the goods market looks as follows:

$$I = Y - S - A - C \qquad (9.15)$$

Expression (9.15) is a result of Walras Law and can be derived by combining (9.4), (9.5), (9.6), (9.11) and (9.13) for dividends, investment, profits, the household budget constraint and the government budget constraint, respectively.

9.1.2   Linearization

To solve the model, we log-linearize it around a steady state. Appendix 9A derives the log-linearized factor-demand equations. Table 9.1 contains the log-linearized model. Notation is explained at the end of the table. Relative changes in the growing variables are presented as a ratio to the capital stock (which is initially predetermined) and denoted by lower-case symbols. To illustrate, the change in the pollution tax -- which is a growing variable -- is given by $\tilde{t}_P = \tilde{T}_P - \tilde{K}$.

Table 9.1: The model in relative changes

Output

$$\tilde{y} = \delta\tilde{s} + \gamma\tilde{a} + \alpha\gamma\tilde{P} + \eta\tilde{M} \tag{T9.1}$$

Pollution

$$[1 + \alpha(1 - \sigma_Y)]\tilde{P} = \frac{\delta(\sigma_H - \sigma_Y)}{\sigma_H(\delta + \beta)}\tilde{s} - (1 - \sigma_Y)(\tilde{t}_P + \tilde{T}_A) + \sigma_Y(\tilde{R} - \tilde{t}_P) \tag{T9.2}$$

Abatement

$$\tilde{a} = \tilde{P} + \tilde{T}_A + \tilde{t}_P \tag{T9.3}$$

Rate of return to capital

$$\tilde{y} = \frac{\delta(\sigma_H - \sigma_Y)}{\sigma_H(\delta + \beta)}\tilde{s} + (1 - \sigma_Y)\eta\tilde{M} + \sigma_Y[\tilde{R} + \tilde{T}_Y] \tag{T9.4}$$

Dividends

$$w_D\tilde{d} = -w_I\tilde{i} + (1 - T_Y)[\delta\tilde{s} + \gamma\tilde{T}_A - \tilde{T}_Y - \alpha\gamma\tilde{t}_P - \epsilon\eta\tilde{P}] \tag{T9.5}$$

Profits

$$w_\Pi\tilde{\pi} = w_D\tilde{d} + w_I\tilde{i} - Rw_K\tilde{R} \tag{T9.6}$$

Growth rate

$$\tilde{g} \equiv \tilde{i} \tag{T9.7}$$

Household savings

$$\tilde{i} = \frac{R}{R - \theta} \tilde{R} \tag{T9.8}$$

Household budget constraint

$$w_C \tilde{c} = w_\Pi \tilde{\pi} + (1 - \sigma) R w_K \tilde{R} \tag{T9.9}$$

Government budget constraint

$$(1 - T_Y) \tilde{T}_Y + T_Y \tilde{y} + T_P^* \tilde{t}_P + T_P^* \tilde{P} = (1 - T_A) w_A \tilde{T}_A + T_A w_A \tilde{a} + w_S \tilde{s} \tag{T9.10}$$

Environmental quality

$$\tilde{M} = -\epsilon \tilde{P} \tag{T9.11}$$

Goods-market equilibrium (Walras Law)

$$w_I \tilde{i} = \tilde{y} - w_C \tilde{c} - w_A \tilde{a} - w_S \tilde{s} \tag{T9.12}$$

Elasticities

$$\beta = \frac{\frac{\partial Y}{\partial K} K}{Y} \qquad \delta = \frac{\frac{\partial Y}{\partial S} S}{Y} \qquad \gamma = \frac{\frac{\partial Y}{\partial A} A}{Y} \qquad \alpha\gamma = \frac{\frac{\partial Y}{\partial P} P}{Y} \qquad \epsilon = -\frac{\partial m}{\partial P} \frac{P}{m} > 0$$

Taxes

$$\tilde{T}_A = \frac{dT_A}{1 - T_A} \quad , \quad \tilde{T}_Y = \frac{dT_Y}{1 - T_Y} \quad , \quad \tilde{T}_P = \frac{dT_P}{T_P}$$

Shares

$$w_A = A/Y \qquad w_D = D/Y \qquad T_P^* = T_P P/Y \qquad w_\Pi = \Pi/Y$$
$$w_S = S/Y \qquad w_C = C/Y \qquad w_I = I/Y \qquad w_k = K/Y$$
$$w_V = V/Y$$

Relationships between the shares

Dividends $w_D = (1-T_Y) - (1-T_A)w_A - T_P^* - w_I$

                                   $w_D = (R\text{-}g)w_V$ (S1)

Profits $w_\Pi = w_D + w_I - Rw_K = (1-T_Y)(\delta - \alpha\gamma)$ (S2)

Household budget constraint $w_C = w_\Pi + (R\text{-}g)w_K$

                                   $w_C = w_D$ (S3)

Definition of investment $w_I = gw_K$ (S4)

Government budget constraint $w_S + T_A w_A = T_Y + T_P^*$ (S5)

Goods-market equilibrium $w_I = 1 - w_S - w_A - w_C$ (S6)

From the first-order conditions

$$(1 - T_Y)\gamma = (1 - T_A)w_A \tag{E1}$$

$$(1 - T_Y)\alpha\gamma = T_P^* \tag{E2}$$

$$(1 - T_Y)\beta = Rw_K \tag{E3}$$

Expression (T9.2) reveals that a higher relative price of pollution induces substitution from pollution towards other inputs. The strength of the substitution effect towards either abatement or capital depends on the magnitude of σ_Y. In particular, if $\sigma_Y < 1$ (as in section 9.3 below), substitution between the intermediate factors H and N is difficult relative to substitution between A and P. In that case, pollution and abatement are substitutes: a higher price for abatement increases pollution. However, if $\sigma_Y > 1$, then pollution, compared to other inputs, is a poor substitute for abatement. Therefore, a higher relative price for abatement lowers pollution. Pollution can be affected also by public investment. In particular, lower public investment raises pollution if $\sigma_H < \sigma_Y$. In that case, compared to capital, pollution is a better substitute for public investment. The demand for abatement rises with the level of pollution, abatement subsidies and pollution taxes (see (T9.3)).

Relations (T9.7) and (T9.8) reveal that the economy grows faster if and only if the interest rate increases. This is because a higher interest rate is associated with a higher after-tax rate of return to capital, which stimulates savings and investment.[106] According to (T9.9), consumption rises if profits increase. Consumption is affected also by the income and intertemporal substitution effects associated with changes in the interest rate. In particular, if the intertemporal elasticity of substitution (σ) exceeds unity, the intertemporal substitution effect dominates and consumption declines on account of a higher interest rate.

9.1.3   Welfare

The welfare effects of small policy changes are measured by the marginal excess burden (*MEB*), which amounts to (see appendix 9B):

$$
MEB = - \left[\frac{\alpha \gamma - \epsilon \eta}{w_C} - \epsilon \phi \right] \tilde{P} - \frac{T_Y - T_A}{1 - T_A} \frac{\gamma}{w_C} \left[\tilde{a} + \frac{g}{R - g} \tilde{g} \right]
$$
$$
- \frac{\delta - w_S}{w_C} \left[\tilde{s} + \frac{g}{R - g} \tilde{g} \right] - \frac{T_Y \beta}{w_C} \frac{g}{R - g} \tilde{g}
$$

(9.16)

All coefficients on the RHS of (9.16) represent wedges between the marginal social costs and the marginal social benefits of an activity. If benefits exceed costs at the margin, an increase in such activity enhances welfare. For the growing variables *A, S* and *K*, expression (9.16) reveals that not only changes in current levels matter for welfare, but also the present value of future changes (indicated by the change in the growth rate).

---

[106] To ensure that growth is positive in the initial equilibrium, we assume that the interest rate exceeds the pure rate of time preference in the initial equilibrium (see (9.12)). Since we consider only marginal changes in the tax system, growth remains positive after the tax reform. Hence, if irreversible investments would restrict the growth rate to be non-negative, a boundary solution with zero growth could not occur. For an analysis of a model with negative growth, see Smulders and Gradus (1996). With negative growth, utility is declining over time. According to the terminology of Pezzey (1992), this implies that the economy is not sustainable.

The first term on the RHS of (9.16) stands for the welfare effect of a change in pollution, which is a non-growing variable. First, as a private input into production, pollution exerts a positive effect on output, which is represented by $\alpha\gamma$. Second, by worsening environmental quality, pollution imposes an adverse production externality, captured by $\epsilon\eta$. The overall effect on productivity is positive only if $\alpha\gamma - \epsilon\eta > 0$ (i.e. if the positive output effect of additional pollution as a rival input for the individual firm exceeds the negative external (non-rival) productivity effect of aggregate pollution). Finally, pollution yields an adverse consumption externality (see the last term between the first square brackets).

The second term on the RHS of (9.16) reveals that higher current or future abatement improves welfare if the output tax, which acts as an implicit tax on abatement, exceeds the abatement subsidy. Hence, more abatement boosts welfare only if it is taxed on a net basis.

The partial welfare effect of higher public investment is captured by the third term on the RHS of (9.16). A rise in public investment improves welfare if the productivity effect of public investment exceeds the share of public investment in output. This condition implies that the marginal product of public investment is larger than unity, i.e. $dY/dS > 1$ or, alternatively, that the marginal social benefits, dY, exceed the marginal social costs, dS.

The last term on the RHS of (9.16) represents the welfare impact of marginal changes in the rates of capital accumulation. The tax on output drives a wedge between the before-tax rate of return to capital (which stands for the marginal social benefits of capital accumulation) and the after-tax rate of return to capital (which measures the marginal social costs of capital accumulation) so that the growth rate of capital is too low from a social point of view. Accordingly, a higher rate of capital accumulation (i.e. $\tilde{g} > 0$) improves welfare because the benefits in terms of higher future output exceed the costs in terms of current consumption foregone.

The double dividend discussion has focused on tax distortions rather than spending distortions. Hence, we abstract from distortions associated with inefficient levels of public spending on infrastructure and abatement subsidies by setting $T_A = T_Y$ and $\delta = w_S$. Changes in abatement or public investment thus yield no first-order effects on welfare. This allows us to concentrate on the interaction between the environmental distortion (i.e. the first term on the RHS of (9.16)) and the tax wedge distorting growth (i.e. the fourth term on the RHS of (9.16)). In particular, an environmental tax reform yields a double dividend if it not only cuts pollution but also alleviates the tax distortion by boosting growth.[107]

9.2   Effects on growth, pollution and welfare

This section explores the economic, ecological and welfare effects of higher pollution taxes (i.e. $\tilde{t}_p > 0$). Appendix 9C solves the model from Table 9.1 for the special case that abatement subsidies and the share of public investment spending in the capital stock remain constant (i.e. $\tilde{T}_A = \tilde{s} = 0$). This section discusses the results for a sub-production function $f(N,H)$ of the Cobb-Douglas type (i.e. $\sigma_Y = 1$),[108] while section 9.3 examines the more general case in which σ_Y is allowed to differ from unity.

9.2.1   Effects on growth and environment

The reduced-form equations for the growth rate, pollution and the tax rate on output are given by:

$$\tilde{g} = -\frac{R}{R-\theta}\frac{1}{\Delta}(\alpha\gamma - \epsilon\eta)\tilde{t}_P \qquad (9.17)$$

---

[107] If $\delta > w_S$ and $T_Y > T_A$, growth would boost welfare not only by alleviating the tax distortion but also by reducing distortions due to inefficient levels of expenditure on infrastructure and abatement subsidies. In that case, we could talk about triple or quadruple dividends. In several papers, Bovenberg and van der Ploeg (1994ab) explore the optimal choice between using pollution taxes and public abatement activities as a response to increasing environmental concern.

[108] The value of σ_H does not affect the analysis because the ratio S/K is fixed. Hence, the sub-production function $h(K,S)$ does not need to be Cobb Douglas.

$$\tilde{P} = -\tilde{t}_P - \frac{1}{\Delta}(\alpha\gamma - \epsilon\eta)\tilde{t}_P \tag{9.18}$$

$$\tilde{T}_Y = \frac{w_S}{\Delta}(\alpha\gamma - \epsilon\eta)\tilde{t}_P \tag{9.19}$$

where $\Delta \equiv Rw_K + w_{\Pi} - (\alpha\gamma - \epsilon\eta) > 0$.[109] Expression (9.18) reveals that a pollution tax exerts two effects on pollution. First, pollution decreases proportionally due to input substitution (i.e. the first term on the RHS of (9.18)). Hence, the *composition* of economic activity becomes cleaner. Second, changes in the growth rate affect pollution (see (9.17) and (9.18)). On account of this growth effect, the *level* of economic activity changes. The overall effect on pollution is unambiguously negative.

The growth effect of an environmental tax reform depends on how such a reform impacts the after-tax rate of return, which determines the incentives to save according to the Keynes-Ramsey rule (9.12). The after-tax rate of return, in turn, is determined by the before-tax rate of of return (i.e. the marginal productivity of capital) and the income tax rate. The net impact on the before-tax rate of return depends on two offsetting effects. On the one hand, a higher pollution tax rate depresses the marginal productivity of capital by reducing the input of pollution. These costs of the pollution tax depend on the production elasticity of pollution, $\alpha\gamma$, which measures the importance of pollution as a private input into production. On the other hand, the improvement in environmental quality associated with the drop in pollution enhances the productivity of inputs in production. These public benefits of a lower level of pollution increase with the environmental externality in production, η, and the elasticity that measures the effect of pollution on environmental quality, ϵ.

If $\alpha\gamma - \epsilon\eta = 0$, a higher pollution tax leaves the marginal product of capital unaffected, as the private costs of less polluting inputs (represented by

---

[109] The determinant, Δ, can be interpreted as the fraction of national income that flows to households. The first term indicates net income from the privately owned capital stock and firm profits. The second term amounts to the net benefits from a lower level of pollution. The determinant should be positive for the equilibrium to be stable.

$\alpha\gamma$) exactly offset the public benefits of a higher level of public environmental services into production (represented by $\epsilon\eta$). In this case, not only the before-tax rate of return but also the after-tax rate of return and hence the growth rate (see (9.17)) remain constant because the tax rate on output, T_y, is unaffected (see (9.19)). The distortionary tax rate must be kept constant because the higher pollution tax does not yield any additional public revenues: with a unitary substitution elasticity, σ_Y, the loss of revenue due to the erosion of the base of the pollution tax exactly matches the increase in revenues from the pollution tax due to the increase in the tax rate. Hence, the slope of the Laffer curve for the pollution tax is zero.

The environmental tax reform harms growth if $\alpha\gamma - \epsilon\eta > 0$. In this case, the after-tax rate of return declines because of two reasons: a decrease in the before-tax rate of return and an increase in the tax rate on output. The before-tax rate of return to capital falls because the adverse input effect of lower pollution on the marginal productivity of capital dominates the favorable externality effect. The drop in the rate of return to capital erodes the base of the output tax. Accordingly, with revenues from the pollution tax remaining constant, the government is forced to raise the tax rate on output in order to meet its revenue constraint.

If $\alpha\gamma - \epsilon\eta < 0$, higher pollution taxes raise the before-tax rate of return. The broadening of the base of the output tax allows for a reduction in the tax rate on output (see (9.19)), thereby ensuring that the after-tax rate of return and thus growth rise. Hence, if the environmental externality in production is large, an environmental tax reform may boost not only the quality of the environment but also growth.

9.2.2   Welfare effects

Armed with the economic and ecological effects, we now turn to the effects on welfare, defined by the marginal excess burden in (9.16). The welfare effects of pollution taxes depend on the effects on only pollution and growth (see (9.16), with $T_A = T_Y$ and $\delta = w_S$). By substituting the reduced-form equations for growth and pollution, (9.17) and (9.18), into (9.16), we derive the following solution for the *MEB*:

$$MEB = -\frac{Rw_K + w_\Pi}{\Delta}\epsilon\phi\tilde{t}_P + [Rw_K + w_\Pi + \frac{g}{R-g}\frac{R}{R-\theta}T_Y\beta]\frac{1}{\Delta}\frac{\alpha\gamma - \epsilon\eta}{w_C}\tilde{t}_P \quad (9.20)$$

The first term on the RHS of (9.20) reveals that, by cutting the level of pollution, pollution taxes raise welfare by enhancing environmental amenities. Second, if a lower pollution level raises the after-tax rate of return to capital (i.e. $\alpha\gamma - \epsilon\eta < 0$), then pollution taxes enhance private or non-environmental welfare. This private component is captured by the second term on the RHS of (9.20). The coefficient between square brackets on the RHS of (9.20) measures this private component. It reveals that a higher after-tax rate of return to capital boosts welfare by raising (current) household income from capital and profits (first term between square brackets) and the growth rate. The welfare impact of changes in the growth rate depends on the wedge between the marginal social costs and benefits of capital accumulation, implied by the distortionary tax on output.

   If higher taxes on pollution raise the after-tax rate of return to capital, then they yield a double dividend by enhancing not only environmental amenities but also productivity and growth and thus private welfare. If, however, the production externality ($\epsilon\eta$) is smaller than the production elasticity ($\alpha\gamma$), then higher pollution taxes reduce the after-tax rate of return to capital. In that case, a trade-off exists between higher environmental amenities and lower economic growth and private welfare.

9.2.3   Optimal pollution taxes

The pollution tax is set at its optimal level if a marginal change in its rate leaves welfare unaffected, i.e. if the marginal excess burden is zero. In a first-best world where the government has access to lump-sum taxes, the optimal pollution tax is at the Pigovian level, which fully internalizes the environmental externalities in production and consumption (see (9.16), with $T_Y = 0$, $T_A = 0$ and $\delta = w_S$):

$$\epsilon \, \varphi \; - \; \frac{\alpha \gamma \; - \; \epsilon \eta}{w_c} \; = \; 0 \tag{9.21}$$

However, our model is not first-best. In particular, the government does not have access to lump-sum taxation to finance its public goods. To investigate whether, in this second-best world, the government still finds it optimal to set the environmental tax at its Pigovian level, we substitute (9.21) into (9.16) and arrive at:

$$MEB \; = \; - \frac{T_Y \beta}{w_c} \, \frac{g}{R-g} \, \tilde{g} \tag{9.22}$$

Hence, if $T_Y > 0$, raising the pollution tax above its Pigovian level harms welfare if it reduces the growth rate. Indeed, at the Pigovian tax the production elasticity of pollution exceeds the environmental externality in production if environmental amenities are positive (i.e. $(\alpha \gamma \; - \; \epsilon \eta)/s_c \; = \; \epsilon \varphi \; > 0$, see (9.21)).

Hence, a higher pollution tax reduces the growth rate (see (9.17)). Intuitively, by internalizing the consumption externality of pollution, environmental taxes reduce the productivity of capital and, therefore, harm growth. This drop in growth produces a first-order loss in welfare, as capital accumulation features a gap between its marginal social costs and marginal benefits.

These results demonstrate that pollution taxes aimed at internalizing environmental externalities may, as a side effect, exacerbate pre-existing distortions in the economy. Indeed, in a second-best world, the optimal tax on pollution differs from its first-best level. In particular, the optimal pollution tax lies below the Pigovian level. This conclusion is consistent with chapter 3 where we explore the interaction between environmental and labor-market distortions.

9.3   Tax shifting

Section 9.2 assumes a unitary substitution elasticity σ_Y, implying that substitution from pollution towards physical capital is relatively easy. In practice, environmental taxes with the most revenue potential involve energy use. Most empirical evidence suggests that substitution possibilities between energy inputs and capital are rather limited (see the references in chapter 4). In this section, therefore, we allow the substitution elasticity σ_Y to be smaller than unity. Accordingly, pollution taxes imply less factor substitution and are thus more effective as a revenue-raising device. Moreover, compared to capital, abatement is a better substitute for pollution. This does better justice to the term abatement.

9.3.1   Effects on growth and pollution

The reduced-form equation for pollution amounts to (see appendix 9C):

$$\tilde{P} = -\left[1 - \frac{(1-\sigma_Y)(1-T_Y)(1-\gamma)\alpha}{\Delta^*} \right]\tilde{t}_p - \frac{\sigma_Y(\alpha\gamma - \epsilon\eta)}{\Delta^*}\tilde{t}_P \qquad (9.23)$$

where $\Delta^* \equiv \sigma_Y[Rw_K + w_\Pi - (\alpha\gamma - \epsilon\eta)] + (1-\sigma_Y)(1-T_Y)(1-\gamma)(1+\alpha) > 0$. If σ_Y equals unity, (9.23) is identical to (9.18). If substitution between the intermediate factors H and N becomes more difficult, however, pollution taxes are less effective in cutting pollution, as they induce less substitution from pollution towards capital. Nevertheless, relation (9.23) reveals that pollution taxes always reduce pollution, even if σ_Y approaches zero.[110]

The reduced form for the tax rate on output is given by:

$$\tilde{T}_Y = \frac{w_S}{\Delta^*}(\alpha\gamma - \epsilon\eta)\tilde{t}_P - \frac{(1-\sigma_Y)(1-T_Y)(\alpha - \epsilon\eta)\alpha\gamma}{\Delta^*}\tilde{t}_P \qquad (9.24)$$

---

[110] If $\sigma_Y = 0$, the coefficient in (9.23) can be rewritten as $-(1-T_Y)(1-\gamma)/\Delta^* < 0$.

This relationship reveals that if substitution between pollution and capital becomes more difficult (i.e. σ_Y becomes smaller), pollution taxes are more efficient as a revenue-raising device. Indeed, even if pollution taxes involve a negative effect on gross productivity (i.e. $\alpha\gamma - \epsilon\eta > 0$), they may nevertheless raise a positive amount of public revenues (which are used to cut the tax rate on output).

The effect on the growth rate is given by:

$$\frac{1}{\sigma} w_I \tilde{g} = (\alpha\gamma - \epsilon\eta)\tilde{P} - w_\Pi \tilde{\pi} \tag{9.25}$$

This expression reveals that the growth performance is affected by two factors, namely, first, the net costs of environmental policy borne by the production sector (i.e. the first term on the RHS of (9.25)), and, second, the part of the costs borne by profits (i.e. the second term on the RHS of (9.25)). As explained in section 9.2, a tighter environmental policy generates an adverse impact on growth if a lower pollution level implies a net burden on the production sector. This is the case if $\alpha\gamma - \epsilon\eta > 0$, i.e. if the negative output effect of a reduction in pollution as a private rival input exceeds the positive effects associated with less adverse production externalities. The second term on the RHS of (9.25) indicates the effect of shifting the tax burden away from the after-tax rate of return towards profits. In particular, lower profits produce a higher growth rate. Intuitively, taxing away a larger share of profit income creates room to cut the distortionary tax rate on output (see (9.24)), thereby raising the return on capital accumulation.

The reduced-form equation for profits is given by:

$$w_\Pi \tilde{\pi} = -\frac{w_\Pi}{\Delta^*}(\alpha\gamma - \epsilon\eta)\tilde{i}_P - \frac{(1-\sigma_Y)(1-T_Y)[\alpha\gamma - \epsilon\eta + Rw_K\alpha]}{\Delta^*}\alpha\gamma\,\tilde{i}_P \tag{9.26}$$

The first term on the RHS of (9.26) represents the effect of input substitution. It reveals that, through this channel, profits share in the burden of environmental policy on the production sector. In particular, pollution taxes harm profits if the adverse pollution externalities on production are small compared to the input share of pollution as a rival input (i.e. $\alpha\gamma - \epsilon\eta > 0$). The second term on the RHS of (9.26) shows the effect of shifting the tax burden away from the after-tax rate of return on investment towards profits. If $\sigma_Y < 1$, the tax-shifting effect causes the replacement of output taxes by pollution taxes to raise the return on investment by reducing profits, thereby boosting growth. The overall effect on growth is positive if the aggregate burden of environmental policy on the production sector is small (i.e. $(\alpha\gamma - \epsilon\eta)\tilde{P}$ is small) and a large share of this burden can be shifted unto profits (i.e. $\tilde{\pi}$ is large and negative). By substituting (9.23) and (9.26) into (9.25), we derive the following condition for pollution taxes to boost the growth rate:

$$\frac{\alpha\gamma - \epsilon\eta}{\alpha} < (1-\sigma_Y)T_P^* \tag{9.27}$$

The LHS of (9.27) represents the net burden of less pollution on the production sector. In particular, if $\alpha\gamma - \epsilon\eta > 0$, a decline in pollution hurts the production sector. The parameter α determines how effective pollution taxes are in cutting pollution. In particular, if α is large, pollution is an important factor of production relative to abatement (i.e. $\alpha\gamma$ is large relative to γ). In that case, substantially lowering pollution is not attractive for firms, as the marginal product of pollution is highly sensitive to the flow of pollution.[111]

The RHS of (9.27) shows the benefits of pollution taxes in terms of shifting the tax burden away from the return on investment towards profits. If $\sigma_Y = 1$, the pollution tax and output tax imply the same relative burden on profits and the return on capital accumulation. Hence, an ecological tax reform does not imply a shift toward non-distortionary lump-sum taxation of profits. Accordingly, the tax-shifting effect exerts no positive effect on growth. If $\sigma_Y < 1$, however, the pollution tax, compared to the output tax, is more effective in raising revenues without depressing the return to investment. The reason is that the tax on

---

[111] In the extreme case that α approaches infinity, pollution is not affected by a rise in the pollution tax. Hence, the costs of pollution taxes, represented by the LHS of (9.27), are zero.

pollution is a more effective instrument to tax away profits (and the quasi-rents from pollution, in particular). Thus, compared to an output tax, a pollution tax incorporates a more important lump-sum element. Accordingly, by using the revenues from the pollution tax to cut the distortionary tax on output (see (9.24)), the government shifts the tax burden away from the return on investment towards profits, thereby boosting growth.

On balance, an environmental tax reform stimulates growth if the positive effect of tax shifting (which allows for a reduction in the wedge between the before- and after-tax rate of return to capital) dominates the negative effect of input substitution on gross productivity associated with more public consumption in the form of more environmental amenities. Chapter 4 also finds that if an environmental tax reform shifts the burden of taxation from labor towards the rents from a fixed factor in production, it may boost employment.

9.3.2   Welfare effects

Section 9.2 finds that, in the Cobb-Douglas case, the optimal pollution tax is always below the Pigovian tax in the presence of environmental amenities. The reason is that lowering the pollution tax from the Pigovian level always boosts growth and, therefore, enhances welfare. If substitution is more difficult, then the welfare effects of raising pollution taxes from its Pigovian level continues to be represented by (9.22). Hence, if pollution taxes succeed in raising growth, they improve welfare. Contrary to the Cobb-Douglas case, pollution taxes may indeed raise the growth rate if $\sigma_Y < 1$, even if environmental amenities cause the production elasticity to exceed the production externality (i.e. if $\alpha\gamma - \varepsilon\eta = \varepsilon\varphi w_C > 0$; see (9.27)). This is because raising pollution taxes may improve the efficiency of taxation by shifting the tax burden from the return on investment towards profits.

The smaller the environmental amenities are, the more likely it will be that a rise in pollution taxes from the Pigovian level stimulates growth and thus improves welfare. This can be seen by substituting the Pigovian rule (9.21) and relation (E2) at the end of Table 9.1 to eliminate $T_p{}^*$ and $\alpha\gamma$ from (9.27):

$$w_C \in \phi [1 - \alpha(1 - \sigma_Y)(1 - T_Y)] < \alpha(1 - \sigma_Y)(1 - T_Y)\in \eta \qquad (9.28)$$

With σ_Y below unity, this inequality for growth to increase is met either if environmental amenities are small compared to production externalities or if large values for α ensure that pollution does not fall much.

9.4   Pollution permits

This section explores the consequences for growth if the government adopts environmental policy instruments that do not charge a price for pollution. In particular, we assume that the government exogenously determines the economy-wide level of pollution, P, by freely providing pollution permits to existing firms.[112]

Appendix 9D solves the model if the level of pollution (i.e. the number of pollution permits) is exogenous. It reveals that expression (9.25) for the growth rate continues to hold. Hence, a tighter environmental policy (i.e. $\tilde{P} < 0$) reduces the growth rate by depressing the after-tax rate of return to investment if two conditions are met. First, environmental policy should impose a net burden on the production sector (i.e. $\alpha\gamma - \in\eta > 0$). Second, this burden cannot be shifted to profits. With freely issued pollution permits, the reduced form for profits is given by:

$$w_\Pi \tilde{\pi} = \frac{\delta}{\beta + \delta}(\alpha\gamma - \in\eta)\tilde{P} \qquad (9.29)$$

---

[112] It is essential to assume that only existing firms receive the permits. Indeed, if also new firms would get free permits, there would be an incentive for existing firms to split up, thereby raising the number of free pollution permits. Sustainability can thus no longer be guaranteed. According to Fullerton and Metcalf (1996), our assumption causes environmental policy to create scarcity rents that form a barrier for new firms to enter the market. Indeed, in our model private firms reap the scarcity rents (or quasi rents) of environmental policy if it boosts profits. This section derives the conditions under which environmental regulation indeed taxes or creates quasi rents that flow to private firms.

Expression (9.29) reveals that a reduction of the number of pollution permits (i.e. $\tilde{P} < 0$) reduces profits if and only if environmental policy imposes a net burden on the production sector (i.e. $\alpha\gamma - \epsilon\eta > 0$). This condition differs from the corresponding condition for the case of pollution taxes (i.e. expression (9.26)) if the substitution elasticity, σ_Y, differs from unity. In particular, pollution taxes are more likely to reduce profits than are freely issued pollution permits if $\sigma_Y < 1$. Intuitively, if firms can not easily replace pollution by other inputs, a large share of the burden of pollution taxes is borne by profits. Hence, pollution taxes succeed in shifting the tax burden away from the after-tax rate of return to investment towards profits.

The reduced-form equation for the growth rate is found by substituting (9.29) into (9.25):

$$\tilde{g} = \frac{R}{R - \theta} \frac{\alpha\gamma - \epsilon\eta}{(1 - T_Y)(1 - \gamma)} \tilde{P} \tag{9.30}$$

Expression (9.30) shows that a double-dividend (i.e. not only a lower level of pollution but also a higher rate of growth) occurs only if a tighter environmental policy confers a net benefit to the production sector, i.e. if the fall in gross productivity on account of less pollution as a rival input into production ($\alpha\gamma$) is dominated by the boost in the gross productivity of capital on account of fewer adverse production externalities ($\epsilon\eta$). This condition generally differs from the corresponding condition (9.27) for the case in which the government adopts pollution taxes rather than freely issued pollution permits to cut pollution. The reason is that, in contrast to freely issued pollution permits, pollution taxes may shift the tax burden between profits and the after-tax rate of return on investment.

This difference between the growth performance of alternative instruments of environmental policy reveals that the distribution of property rights of the environment may affect not only the distribution of income but also growth. In particular, if the government owns the property rights of the environment, it can adopt instruments that raise public revenues from the sale of the publicly owned environment, e.g. environmental taxes or auctioned pollution permits. However, if the property rights would be assigned to polluters, then environmental policy instruments should be adopted that do not raise public revenues, such as pollution standards or freely distributed pollution permits.

The ease with which other inputs can substitute for pollution determines whether environmental policy instruments that change the price for pollution are more likely to enhance growth than those instruments that do not. In particular, comparison of (9.27) and (9.30) reveals that, compared to freely issued pollution permits, pollution taxes are more likely to enhance growth if $\sigma_Y < 1$. Intuitively, if substitution possibilities are limited, pollution constitutes a relatively inelastic tax base. Hence, pollution taxes are an attractive instrument to raise revenues because profits, rather than the after-tax rate of return to capital accumulation, bear a large part of the burden of these taxes.[113]

If pollution can be easily replaced by other inputs (i.e. $\sigma_Y > 1$), in contrast, freely issued pollution permits are more likely to improve growth than pollution taxes are. The reason is that, with easy substitution, the base of the pollution tax is relatively elastic. Accordingly, the pollution tax is a relatively inefficient device to raise revenues. In particular, the burden of the pollution tax is reflected mainly in a lower after-tax rate of return to investment rather than in a lower profit level. Thus, the revenues generated by the pollution tax are relatively expensive in terms of a lower after-tax rate of return on investment. Indeed, for any particular amount of revenues, and given the environmental objective, the pollution tax hurts the after-tax rate of return to capital more than the output tax because the pollution tax is less effective in taxing away profits.

The optimal level of pollution permits is found by substituting (9.30) into (9.16) and setting $MEB = 0$:

$$[1 + \frac{r}{R-\theta} \frac{g}{R-g} \frac{T_Y}{1-T_Y} \frac{\beta}{\beta+\delta}] \frac{\alpha\gamma - \epsilon\eta}{w_C} - \epsilon\phi = 0 \qquad (9.31)$$

At the optimal level of pollution permits, the environmental benefits of a marginal reduction in pollution exceed private abatement costs if the

---

[113] See also Nielsen, Pedersen and Sørensen (1995).

environment yields positive environmental amenities (i.e. $\dfrac{\alpha \gamma - \epsilon \eta}{w_C} - \epsilon \phi < 0$).

Intuitively, in this second-best world, increasing the environmental distortion (by raising pollution above the level where environmental benefits equal abatement costs) reduces the distortion arising from distortionary taxation ($T_Y > 0$). In particular, distortionary taxation implies that growth is 'too low:' an increase in growth yields a first-order improvement in welfare. If environmental damages of increased pollution equal private costs of pollution abatement, then the change in environmental quality yields only second-order welfare effects. However, an increase in pollution boosts growth (if environmental amenities cause the private costs of pollution abatement to exceed the positive productivity effects of less pollution). Accordingly, a marginal increase in pollution above the level at which marginal environmental costs equal the marginal reduction in abatement costs boosts welfare: it yields a first-order gain as a result of the growth wedge, but only a second-order loss as a result of the environmental wedge.

9.5   Conclusions

This chapter explores the consequences of an environmental tax reform for economic growth, pollution and welfare in a second-best dynamic framework. It has uncovered two channels through which an environmental tax reform may yield a double dividend -- not only improving environmental quality, but also boosting non-environmental welfare by stimulating economic growth. The first channel is a positive environmental externality in production associated with the role of the environment as a public production factor. The second channel through which an environmental tax reform may boost growth is a shift in the tax burden away from the return on capital accumulation towards profits. In particular, a pollution tax may improve the efficiency of the tax system as a reve-nue-raising device by taxing away the quasi-rents from pollution. This second channel operates only if substitution between pollution and other inputs is difficult -- so that the base of the pollution tax is rather inelastic.

This chapter finds that the role of the environment as a public production factor does not overturn the result derived in chapter 3 that the optimal pollution tax is lower than the Pigovian tax. As long as the environment produces some environmental amenities as a public consumption good, then the Pigovian tax exceeds the optimal pollution tax, which internalizes the adverse pollution externalities on consumption. Intuitively, at the optimal tax level, the pollution tax reduces growth, thereby worsening distortions due to the absence of lump-sum taxation. However, we also demonstrated that the optimal environmental tax may exceed its Pigovian level if the government has no access to more direct ways to tax rents. Intuitively, in the absence of an explicit lump-sum tax on profits, the pollution tax acts an implicit way to tax rents. For the optimal environmental tax to exceed the Pigovian level, the pollution tax should be rather ineffective in cutting pollution. In that case, the shifting of the burden of public spending and environmental amenities to the owners of the firm offsets the costs of additional environmental amenities, thereby alleviating the net burden on investors so that the after-tax rate of return on investment, and with it growth, rise.

Compared to environmental policy instruments that do not charge a price for pollution, pollution taxes may be more efficient instruments to cut pollution. In particular, we find that, if substitution between pollution and other inputs is difficult, pollution taxes are relatively efficient instruments to raise revenue. The intuition is that, with poor substitution, pollution taxes are relatively efficient instruments to raise revenue. Accordingly, they reduce the after-tax rate of return to capital less and profits more than taxes on output do. However, if substitution between pollution and other inputs is easy, then freely issued pollution permits protect the after-tax rate of return to capital better than pollution taxes do.

Appendix 9A   Linearizing the factor-demand equations

Specification of F

$F = f(H,N)$ is homogenous of degree 1 in H and N. Hence, $F/H = f(1,N/H) \to \phi$ $= \phi(z)$ where $\phi = F/H$ and $z = N/H$. The first-order derivatives of F are:

$$\frac{\partial F}{\partial N} = \phi' \tag{9A.1}$$

$$\frac{\partial F}{\partial H} = \phi - z\phi' \tag{9A.2}$$

The substitution elasticity between H and N is defined as:

$$\frac{1}{\sigma_Y} = - \frac{\partial\left(\dfrac{\partial F/\partial N}{\partial F/\partial H}\right)}{\partial\dfrac{N}{H}} \frac{\dfrac{N}{H}}{\dfrac{\partial F/\partial N}{\partial F/\partial H}} = \frac{-z\phi''}{\phi'}\frac{\phi}{\phi-z\phi'} \tag{9A.3}$$

Specification of M

$H = h(K,S)$ is homogenous of degree 1 in K and S. Hence, $H/K = h(1, S/K) \to \mu$ $= \mu(s)$, where $\mu = H/K$ and $s = S/K$. The first-order derivatives of H are:

$$\frac{\partial H}{\partial S} = \mu' \tag{9A.4}$$

$$\frac{\partial H}{\partial K} = \mu - s\mu' \tag{9A.5}$$

The substitution elasticity between K and S is defined as:

$$\frac{1}{\sigma_H} = - \frac{\partial\left(\dfrac{\partial H/\partial S}{\partial H/\partial K}\right)}{\partial\dfrac{S}{K}} \frac{\dfrac{S}{K}}{\dfrac{\partial H/\partial S}{\partial H/\partial K}} = \frac{-s\mu''}{\mu'}\frac{\mu}{\mu-s\mu'} \tag{9A.6}$$

From the definitions of the production elasticities, the Cobb-Douglas specification of N and (9A.4) and (9A.5), we derive:

$$\gamma = \frac{\partial Y}{\partial A}\frac{A}{Y} = [\frac{\partial F}{\partial N}\frac{N}{F}][\frac{\partial N}{\partial A}\frac{A}{N}] = \frac{\partial F}{\partial N}\frac{N}{F} * 1 \qquad (9A.7)$$

$$\beta + \delta = \frac{\partial Y}{\partial K}\frac{K}{Y} + \frac{\partial Y}{\partial S}\frac{S}{Y} = \frac{\partial F}{\partial H}\frac{H}{F}[\frac{\partial H}{\partial K}\frac{K}{H} + \frac{\partial H}{\partial S}\frac{S}{H}] = \frac{\partial F}{\partial H}\frac{H}{F} * 1 \qquad (9A.8)$$

Homogeneity of F and H, and use of (9A.7) and (9A.8) implies:

$$\tilde{F} = \gamma\tilde{N} + (\delta + \beta)\tilde{H} = \tilde{N} + (\delta + \beta)[\tilde{H} - \tilde{N}] \qquad (9A.9)$$

$$\tilde{H} = \frac{\beta}{\beta + \delta}\tilde{K} + \frac{\delta}{\beta + \delta}\tilde{S} \qquad (9A.10)$$

where we have used (9.2).

Specification of N and Y
The Cobb-Douglas specification for N implies for the relative change in N:

$$\tilde{N} = \tilde{A} + \alpha\tilde{P} \qquad (9A.11)$$

Moreover, the externality of pollution specified in (9.1) implies for the relative change in Y:

$$\tilde{Y} = \tilde{F} + \eta\tilde{M} \qquad (9A.12)$$

Linearizing marginal factor productivity
Armed with this set of definitions and relations, we are able to log-linearize the factor-demand equations (9.7), (9.8) and (9.9) for abatement, pollution and capital, respectively. The relative change in the marginal product of abatement is derived by using (9A.1), (9A.3) and the Cobb-Douglas specification of N in the production function (9.1):

$$\frac{\Delta \partial Y / \partial A}{\partial Y / \partial A} = \frac{\Delta \partial F / \partial N}{\partial F / \partial N} + \eta \tilde{M} + \frac{\Delta \partial N / \partial A}{\partial N / \partial A} =$$

$$- \frac{1}{\sigma_Y} (\beta + \delta) [\tilde{N} - \tilde{H}] + \alpha \tilde{P} + \eta \tilde{M}$$

(9A.13)

The linearized marginal product of pollution is found by using the same relations as for abatement:

$$\frac{\Delta [\partial F / \partial P] M^\eta}{\partial Y / \partial P} = \frac{\Delta \partial F / \partial N}{\partial F / \partial N} + \eta \tilde{M} + \frac{\Delta \partial N / \partial P}{\partial N / \partial P} =$$

$$- \frac{1}{\sigma_Y} (\beta + \delta) [\tilde{N} - \tilde{H}] + \tilde{A} - (1 - \alpha) \tilde{P} + \eta \tilde{M}$$

(9A.14)

To derive an expression for the relative change in the marginal product of capital, we use (9A.2), (9A.3), (9A.5), (9A.6) and the specification of the production function (9.1):

$$\frac{\Delta \partial Y / \partial K}{\partial Y / \partial K} = \frac{\Delta \partial F / \partial H}{\partial F / \partial H} + \eta \tilde{M} + \frac{\Delta \partial H / \partial K}{\partial H / \partial K} -$$

$$\frac{1}{\sigma_Y} \gamma [\tilde{N} - \tilde{H}] + \frac{1}{\sigma_H} \frac{\delta}{\beta + \delta} [\tilde{S} - \tilde{K}] + \eta \tilde{M}$$

(9A.15)

Linearizing factor-demand equations

Using (9A.13), we linearize the demand for abatement (9.7) as:

$$\frac{1}{\sigma_Y} (\beta + \delta) [\tilde{N} - \tilde{H}] - \alpha \tilde{P} - \eta \tilde{M} = \tilde{T}_A - \tilde{T}_Y$$

(9A.16)

From (9A.4), we arrive at the following linearized demand for pollution (9.8):

$$- \frac{1}{\sigma_Y} (\beta + \delta) [\tilde{N} - \tilde{H}] + \tilde{A} - (1 - \alpha) \tilde{P} + \eta \tilde{M} = \tilde{T}_P + \tilde{T}_Y$$

(9A.17)

By adding (9A.16) and (9A.17), we find the following relation between abatement and pollution:

$$\tilde{A} = \tilde{P} + \tilde{T}_P + \tilde{T}_A$$

(9A.18)

Combining (9A.10) and (9A.18), we derive for abatement and pollution:

$$\tilde{A} = \frac{1}{1+\alpha}\tilde{N} + \frac{\alpha}{1+\alpha}[\tilde{T}_P + \tilde{T}_A] \qquad (9A.19)$$

$$\tilde{P} = \frac{1}{1+\alpha}\tilde{N} - \frac{1}{1+\alpha}[\tilde{T}_P + \tilde{T}_A] \qquad (9A.20)$$

Finally, (9A.15) and (9.9) yield for the desired demand for capital:

$$\frac{1}{\sigma_Y}\gamma[\tilde{N}-\tilde{H}] + \frac{1}{\sigma_M}\frac{\delta}{\beta+\delta}[\tilde{S}-\tilde{K}] + \eta\tilde{M} = \tilde{R} + \tilde{T}_Y \qquad (A21)$$

Now we are able to derive the factor-demand equations as presented in table 9.1. First, the rate of return to capital (T9.4) is found by substituting (9A.9), (9A.10) and (9A.12) into (9A.21) to eliminate $\tilde{N}$, $\tilde{H}$ and $\tilde{F}$, respectively. Second, relation (T9.2) for pollution is found by substituting (9A.10) and (9A.11) into (9A.16) to eliminate $\tilde{N}$ and $\tilde{H}$, and then using (T9.4) to eliminate $\tilde{Y}$. Finally, (T9.3) is equivalent to (9A.18).

Appendix 9B   Welfare effects

This appendix derives the marginal excess burden as a measure for changes in social welfare. It represents the welfare changes due to various policy measures in terms of consumption.

Utility is defined in (9.10). On the sustainable balanced-growth path, private consumption grows at a constant rate, g, while pollution remains constant. Hence, we can rewrite utility as:

$$U = \frac{[C_0 M^{\phi}]^{1-\frac{1}{\sigma}}}{1-\frac{1}{\sigma}} \int_0^{\infty} e^{[g(1-\frac{1}{\sigma})-\theta]t} \, dt \qquad (9B.1)$$

where C_0 is the initial consumption level. To solve the integral at the RHS of (9B.1), we require that $\pi(1 - 1/\sigma) - \theta < 0$ so that utility is bounded. Using (9.12), we rewrite this condition as follows:

$$\theta - g(1 - \frac{1}{\sigma}) = R - g > 0 \tag{9B.2}$$

Solving the integral in (9B.1) yields the following expression for utility:

$$U = \frac{[C_0 M^\phi]^{1-\frac{1}{\sigma}}}{(1 - \frac{1}{\sigma})(\theta - g(1 - \frac{1}{\sigma}))} \tag{9B.3}$$

For utility to remain constant after a policy shock, we need:

$$0 = dU = \frac{\partial U}{\partial C}dC + \frac{\partial U}{\partial M}dM + \frac{\partial U}{\partial g}dg \tag{9B.4}$$

The derivatives at the RHS of (9B.4) can be derived by differentiating (9B.3) with respect to its arguments. This yields:

$$\frac{\partial U}{\partial C}dC = \frac{[C_0 M^\phi]^{1-\frac{1}{\sigma}}}{R - g}\tilde{C} \tag{9B.5}$$

$$\frac{\partial U}{\partial M}dM = \frac{[C_0 M^\phi]^{1-\frac{1}{\sigma}}}{R - g}\phi\tilde{M} \tag{9B.6}$$

$$\frac{\partial U}{\partial g}dg = \frac{[C_0 M^\phi]^{1-\frac{1}{\sigma}}}{R - g}\frac{g}{R - g}\tilde{g} \tag{9B.7}$$

Substituting (9B.5), (9B.6) and (9B.7) into (9B.4), we arrive at the following expression for the change in welfare:

$$MEB + \tilde{C} + \phi\tilde{M} + \frac{g}{R-g}\tilde{g} = 0 \tag{9B.8}$$

where *MEB* is the additional consumption that needs to be provided to the household in order to keep utility, after the policy shock, at its initial level.

To arrive at an alternative formulation for the *MEB* in (9B.8), we substitute (T9.1) into (T9.12) to eliminate consumption. Thus, we find:

$$MEB = -\phi\tilde{M} - \frac{g}{R-g}\tilde{g} - \frac{\delta - w_S}{w_C}\tilde{s} - \frac{\gamma - w_A}{w_C}\tilde{a} - \frac{\alpha\gamma - \epsilon\eta}{w_C}\tilde{P} + \frac{w_I}{w_C}\tilde{i} \tag{9B.9}$$

To rewrite (9B.9), we first substitute (T9.11) into (9B.9) to eliminate $\tilde{M}$. Then we use the production elasticity (E1) at the end of Table 9.1 to rewrite the coefficient for abatement. By using (T9.7), we can eliminate $\tilde{i}$. Finally, by using the relations between the shares (S2) and (S6), the production elasticity (E3) and relation (9.2), we derive the following relationship:

$$(R-g)w_K = w_C - (\delta - w_S) - (\gamma - w_A) - T_Y\beta \qquad (9B.10)$$

Using (9B.10) and (S2), we can rewrite the coefficient for growth in the *MEB* in (9B.9) as in (9.16).

Appendix 9C   Solution of the model

This appendix solves the model of Table 9.1 for the case in which the government uses T_Y to balance its budget (ex post). We assume that $\delta = w_S$, $T_A = T_Y$ and $\tilde{T}_A = \tilde{s} = 0$. This appendix makes extensive use of the relations between shares and the definitions of the production elasticities at the end of table 9.1.

By substituting (T9.3) and (T9.11) into (T9.1), we can write output as:

$$\tilde{y} = [\gamma + \alpha\gamma - \epsilon\eta]\tilde{P} + \gamma\tilde{t}_P \qquad (9C.1)$$

Then we substitute (9C.1) into (T9.4) to eliminate $\tilde{y}$ and use (T9.11) to eliminate $\tilde{M}$. This yields:

$$\sigma_y\tilde{r} = [\gamma + \alpha\gamma - \sigma_y\epsilon\eta]\tilde{P} + \gamma\tilde{t}_P - \sigma_y\tilde{T}_y \qquad (9C.2)$$

Substitution of (9C.2) into (T9.2) to eliminate $\tilde{R}$, we arrive at the following relation:

$$[(1-\gamma)(1+\alpha) + \sigma_y(\epsilon\eta - \alpha)]\tilde{P} + \sigma_y\tilde{T}_Y = -(1-\gamma)\tilde{t}_P \qquad (9C.3)$$

By substituting $\tilde{y}$ and $\tilde{a}$ from (9C.1) and (T9.3) into the government budget constraint (T9.10), we find another expression for $\tilde{P}$ and $\tilde{T}_Y$ in terms of the exogenous variables:

$$[\alpha \gamma - T_Y \in \eta] \tilde{P} + (1 - T_Y) \tilde{T}_Y = - T_P^* \tilde{t}_P \tag{9C.4}$$

Relations (9C.3) and (9C.4) form two linear equations in two endogenous variables. Rewriting them in matrix notation yields:

$$\begin{pmatrix} 1 - T_Y & \alpha \gamma - T_Y \in \eta \\ \sigma_Y & (1 - \gamma)(1 + \alpha) + \sigma_Y(\in \eta - \alpha) \end{pmatrix} \begin{pmatrix} \tilde{T}_Y \\ \tilde{P} \end{pmatrix} = \begin{pmatrix} - T_P^* \\ -(1 - \gamma) \end{pmatrix} \tilde{t}_P \tag{9C.5}$$

Solving this linear system by inverting the matrix at the LHS of (9C.5), we find:

$$\Delta^* \begin{pmatrix} \tilde{T}_Y \\ \tilde{P} \end{pmatrix} = \begin{pmatrix} (1 - \gamma)(1 + \alpha) + \sigma_Y(\in \eta - \alpha) & T_Y \in \eta - \alpha \gamma \\ - \sigma_Y & 1 - T_Y \end{pmatrix} \begin{pmatrix} - T_P^* \\ -(1 - \gamma) \end{pmatrix} \tilde{t}_P \tag{9C.6}$$

where:

$$\Delta^* = (1 - T_Y)(1 - \gamma)(1 + \alpha) - \sigma_Y[\alpha \gamma - \in \eta + (1 - T_Y)\alpha] \tag{9C.7}$$

$$= \sigma_Y[Rw_K + w_\Pi - (\alpha \gamma - \in \eta)] + (1 - \sigma_Y)(1 - T_Y)(1 \gamma)(1 + u)$$

The determinant in (9C.7) should be positive for the equilibrium to be stable. (9C.6) yields the reduced-form equations for the output tax and pollution:

$$\Delta^* \tilde{T}_Y = w_S(\alpha \gamma - \in \eta) \tilde{t}_P - (1 - \sigma_Y) T_P^* (\alpha - \in \eta) \tilde{t}_P \tag{9C.8}$$

$$\Delta^* \tilde{P} = - [Rw_k + w_\Pi] \tilde{t}_P - (1 - \sigma_Y) T_P^* \tilde{t}_P$$

$$= - \Delta^* \tilde{t}_P + (1 - \sigma_Y)(1 - T_Y)(1 - \gamma)\alpha \tilde{t}_P - \sigma_Y(\alpha \gamma - \in \eta) \tilde{t}_P \tag{9C.9}$$

Substituting (9C.3) into (T9.2) to eliminate $\tilde{P}$, we find for the interest rate:

$$[(1 - \gamma)(1 + \alpha) + \sigma_Y(\in \eta - \alpha)] \tilde{R} = - (\alpha \gamma - \in \eta) \tilde{t}_P - [1 + \alpha(1 - \sigma_Y)] \tilde{T}_Y \tag{9C.10}$$

In order to find the reduced form for the interest rate, we substitute (9C.8) into (9C.10). After some algebra, we find:

$$\Delta^* \tilde{R} = - (\alpha \gamma - \in \eta) \tilde{t}_P + (1 - \sigma_Y)(1 - T_Y) \alpha \gamma \alpha \tilde{t}_P \tag{9C.11}$$

The reduced form for the growth rate is closely related to (9C.11) (see (T9.7) and (T9.8)). The reduced form for the marginal excess burden can be derived by substituting the reduced forms for pollution and growth into (9.16).

In order to gain more insight into the effects of pollution taxes on the interest rate, we derive another expression for $\tilde{R}$. By substituting (T9.5) into (T9.6) to eliminate $\tilde{d}$, we find for profits:

$$w_{\Pi} \tilde{\pi} = -(1 - T_Y)[\tilde{T}_Y + \alpha\gamma \tilde{i}_P + \epsilon\eta \tilde{P} + \beta\tilde{R}] \qquad (9C.12)$$

Substituting (9C.4) into (9C.12) to eliminate $\tilde{T}_Y$, and rearranging terms, we find the following expression for the interest rate:

$$(1 - T_Y)\beta\tilde{R} = (\alpha\gamma - \epsilon\eta)\tilde{P} - w_{\Pi}\tilde{\pi} \qquad (9C.13)$$

The reduced form for profits is derived by substituting (9C.9) and (9C.11) into (9C.13).

$$\Delta^* \tilde{\pi} = -(\alpha\gamma - \epsilon\eta)\tilde{i}_P - \frac{\alpha\gamma}{\delta - \alpha\gamma}(1 - \sigma_Y)[\alpha\gamma - \epsilon\eta + \alpha Rw_K]\tilde{i}_P \qquad (9C.14)$$

Appendix 9D   Pollution permits

This appendix solves the model of Table 9.1 if the level of pollution, P, rather than the pollution tax, T_P, is an exogenous policy variable. Furthermore, T_P does not represent a tax, but rather the shadow price of pollution. Accordingly, the model differs from Table 9.1 as, on the one hand, T_P does not involve costs for firms in the form of tax payments while, on the other hand, it does not yield public revenues. In particular, the term $T_P P$ disappears from equations (9.4) and (9.13) for dividends and the government budget constraint, respectively. Assuming that $\tilde{s} = \tilde{T}_A = 0$ and $T_A = T_Y$, $\delta = w_S$, this involves the following changes for Table 9.1:

Dividends

$$w_D \tilde{d} = -w_I \tilde{i} - (1 - T_Y)[\tilde{T}_Y - (\alpha \gamma - \epsilon \eta) \tilde{P}] \qquad (T9.5')$$

Government budget constraint

$$(1 - T_Y)\tilde{T}_Y = -T_Y \tilde{y} + T_A w_A \tilde{a} = -T_Y(\alpha \gamma - \epsilon \eta) \tilde{P} \qquad (T9.10')$$

where the second expression is found by using (T9.1) and (T9.3). Substitution of (T9.5') into (T9.6), we find for profits:

$$w_\Pi \tilde{\pi} = -(1 - T_Y)[\tilde{T}_Y - (\alpha \gamma - \epsilon \eta) \tilde{P} + \beta \tilde{R}] \qquad (9D.1)$$

Substituting (T9.10') into (9D.1) to eliminate $\tilde{T}_Y$, we arrive at the following expression for $\tilde{R}$:

$$(1 - T_Y)\beta \tilde{R} = (\alpha \gamma - \epsilon \eta) \tilde{P} - w_\Pi \tilde{\pi} \qquad (9D.2)$$

Expression (9D.2) is equivalent to (9C.14) in appendix 9C. Hence, this relationship holds, irrespective of whether pollution taxes or pollution permits are adopted as environmental policy instruments.

In order to find the reduced form for the interest rate, we substitute (T9.3) into (T9.2) to eliminate $\tilde{i}_P$. This yields the following relation between $\tilde{a}$, $\tilde{R}$ and $\tilde{P}$:

$$\sigma_Y \tilde{R} = \alpha(1 - \sigma_Y)\tilde{P} + \tilde{a} \qquad (9D.3)$$

By substituting (T9.10') into (T9.4) to eliminate $\tilde{T}_Y$, and combining this result with (T9.1) to eliminate $\tilde{y}$, we find another expression between abatement, the interest rate, and pollution:

$$\sigma_Y(1 - T_Y)\tilde{R} - (1 - T_Y)\gamma \tilde{a} = [(1 - T_Y)(1 - \sigma_Y)\alpha \gamma + \sigma_Y(\alpha \gamma - \epsilon \eta)]\tilde{P} \quad (9D.4)$$

We now substitute $\tilde{a}$ from (9D.3) into (9D.4) to arrive at the reduced form for the interest rate:

$$\tilde{R} = \frac{\alpha \gamma - \epsilon \eta}{(1 - T_Y)(1 - \gamma)} \tilde{P} \qquad (9D.5)$$

The reduced form for growth is found by means of (T9.7) and (T9.8). Combining (9D.5) with (9D.2), we derive the reduced-form equation for profits:

$$w_{\Pi} \tilde{\pi} = \frac{\delta}{\beta + \delta} (\alpha \gamma - \epsilon \eta) \tilde{P} \qquad\qquad (9D.6)$$

10   Conclusions

This book explores whether a revenue-neutral replacement of taxes on labor for taxes on polluting activities produces a double dividend, i.e. whether it yields environmental benefits and, at the same time, improves welfare from a non-environmental point of view. Using a simple general equilibrium framework, chapter 3 reveals that the double dividend typically fails. Indeed, environmental taxes increase welfare through environmental benefits. However, they typically raise the overall burden of taxation on private incomes, even if the revenues are returned to private agents through reductions in distortionary labor taxes. The intuition is that an ecological tax reform amounts to a shift from a broad-based income tax towards a narrow-based pollution tax. The narrow-based pollution tax aims at encouraging substitution away from polluting behavior. Although benefitting the environment, such behavioral responses reduce tax revenues, which renders these taxes relatively inefficient as instruments to raise public revenue. Indeed, the objective to encourage environmentally-friendly behavior conflicts with the aim of the tax system to raise public revenues with the lowest cost to private incomes.

The model of chapter 3 contains some restrictive assumptions that are unlikely to be met in reality. To make a more realistic assessment of the welfare effects of an environmental tax reform, subsequent chapters extend the simple framework in several directions. These extended models reveal additional channels through which an environmental tax reform affects non-environmental welfare.

Chapter 4 shows that an ecological tax reform may induce a so-called tax shifting effect by shifting the tax burden between factors of production. In particular, we find that a double dividend may be feasible if an environmental tax reform succeeds in shifting the tax burden from labor towards immobile capital. If capital is mobile, however, tax shifting from labor to capital makes a double dividend less likely to occur.

Chapter 5 reveals that a double dividend is feasible if the home country exerts monopsony power on the international market for polluting goods. In that case, environmental levies operate as implicit import tariffs. Indeed, by changing the terms of trade in favor of the home country, environmental taxes shift the burden of taxation towards foreign suppliers of polluting inputs. Although this raises non-environmental welfare in the home country, it involves a loss in welfare of its trading partners.

Chapter 6 illustrates the trade-offs between efficiency, environmental quality and equity. In particular, a reform from labor to pollution taxes may produce a double dividend if the burden of taxation is shifted away from labor incomes towards transfer incomes. However, such a double dividend typically comes at the cost of a less equitable income distribution. If the government features preferences for equity, it can maintain the income distribution by using an alternative revenue-recycling strategy. In that case, however, an ecological tax reform fails to yield a second dividend in the form of more private income.

Chapter 7 explores the opportunities for a double dividend in the presence of labor-market imperfections. It finds that a double dividend is more likely to be obtained if an ecological tax reform increases the income differentials between workers and people outside the labor force. In that case, an environmental tax reform hurts the bargaining position of the union in wage negotiations, thereby reducing wages and stimulating employment. Accordingly, a double dividend becomes feasible. However, this double dividend comes at the cost of a less equitable income distribution, as it favors workers relative to the unemployed.

Chapter 8 incorporates feedback effects of environmental quality on the economy. The analysis reveals that non-environmental welfare may rise due to an environmental tax reform if production externalities, associated with the role of the environment as a public production factor, are sufficiently strong. Furthermore, environmental taxes are more likely to stimulate employment if environmental quality as a public consumption good is a sufficiently strong substitute for leisure.

Table 10.1: Simulations of a 10% tax according to the models of this book.

Model	Relevant details	Revenue recycling	Employment	Private income
Environmental taxes on firms				
Chapter 3	Benchmark	Labor tax	− 0.07	− 0.14
Chapter 4	Mobile capital[a]	Labor tax	− 0.07	− 0.13
		Capital tax	0.16	0.28
	Fixed capital[a]	Labor tax	0.23	0.03
Chapter 5	Monopsony power[b]	Labor tax	0.16	0.31
	Monopoly power[c]	Labor tax	0.18	0.39
Chapter 7	Imperfect labor-market[d]	Labor tax	0.63	0.17
Chapter 8	Production externality[e]	Labor tax	0.13	0.26
Environmental taxes on households				
Chapter 3	Benchmark	Labor tax	− 0.05	− 0.10
Chapter 6	No indexation[f]	Labor tax	0.17	− 0.02
	Indexation[g]	Labor tax; Transfers; TFPA	− 0.05	− 0.10
Chapter 8	Non-separable cons. externality[h]	Labor tax	0.01	− 0.07
	Non-separable cons. externality[i]	Labor tax	− 0.07	− 0.12

[a] Base-run simulation; [b] Endogenous P_E and $\phi = 0.5$; [c] Endogenous P_Y and $\epsilon_{YY} = 2$; [d] $\beta = 1$; $\gamma = 0.5$; $\delta = 2$; [e] $Y_M = 0.1$; [f] Without indexation of non-labor incomes (private income expressed in terms of output); [g] With indexation of real non-labor incomes to real after-tax wage rate (private income expressed in terms of output); [h] Non-separable consumption externality and $\sigma_M = 0.5$; [i] Non-separable consumption externality and $\sigma_M = 4.5$;

Chapter 9 extends the double-dividend analysis to a dynamic context of endogenous growth. This chapter reveals that the main conclusions on the double dividend derived in the static models remain valid in a dynamic framework. Furthermore, it explores under what conditions environmental taxes are more attractive instruments to cut pollution than freely issued pollution permits. It finds that environmental taxes are relatively attractive if pollution features an inelastic tax base.

Chapters 3 - 8 numerically illustrate the main theoretical findings from the various models. In particular, the models are calibrated for a typical small open economy such as the Netherlands. For polluting inputs or polluting consumption commodities, we employ energy data. The behavioral elasticities are based on the available empirical evidence.

Table 10.1 summarizes the effects of an environmental tax reform on employment and private income -- which measure the non-environmental welfare second dividend -- in the various chapters of this book. Table 10.1 reveals that an environmental tax reform reduces employment and private income in the "benchmark" model of chapter 3. Various tax shifting effects may render a double dividend feasible, however. In particular, according to the simulations in chapter 4, employment and private income expand by 0.23% and 0.03%, respectively, if an environmental tax reform moves the tax burden onto fixed capital. In chapter 5, terms-of-trade improvements cause an energy tax reform to raise employment by 0.16% to 0.18%, and private incomes by 0.31% to 0.39%. According to chapters 6 and 7, shifts in the income distribution from labor towards transfer recipients may boost employment by between 0.17% and 0.63%. Chapter 7 reports that private incomes may increase by 0.17%. In chapter 8, strong productivity effects of the environment cause employment and private income to rise by, respectively, 0.13% and 0.26%.

The transmission channels discussed in this book generally play an important role in applied models that have calculated the economic effects of an environmental tax reform. Our analysis may help to understand the simulation results from these applied studies on the double dividend. This is important because empirical models yield a variety of results that are sometimes difficult to reconcile. Indeed, the underlying trade-offs and transmission channels derived from the theoretical models are not always clear from the empirical studies. By combining analytical models, which illustrate the important trade-offs and economic mechanisms, with empirical information on the key parameters, this book bridges the gap between theoretical and applied studies.

How should we interpret the findings from this book? The various extensions of the benchmark model show that, in principle, initial inefficiencies in the tax system may render a double dividend feasible. Indeed, an environmental tax reform can induce a tax shifting effect, which may enhance welfare from a non-environmental point of view. However, the tax shifting effect typically leads to trade-offs. Indeed, one should wonder why governments have not already adopted more direct instruments to alleviate initial inefficiencies. To illustrate, inefficiencies may originate in distributional concerns or in international free-trade agreements that serve their own social objectives. If governments alleviate these inefficiencies under the flag of an environmental tax reform, this will generally damage these other objectives. This typically reduces the probability that consensus will be reached about the introduction of environmental taxes. In general, political considerations call for a minimization of tax shifting effects. Indeed, political viability of environmental taxes is typically strengthened if the government uses the revenues from these taxes to compensate polluters for their income losses. Without a tax shifting effect, however, an environmental tax reform is unlikely to yield a double dividend.

The overall conclusion of this book must be that obtaining a double dividend is unlikely. This conclusion should not be interpreted as an argument against the introduction of environmental taxes. Indeed, these taxes may be attractive instruments to reduce pollution in a cost-effective way, thereby raising social welfare through environmental benefits. Environmental taxes should be justified on the basis of these (possibly considerable) favorable effects on environmental quality, rather than on non-environmental welfare gains.

References

Alfsen, K., A. Brendemoen and S, Glomsrød, 1992, Benefits of climate policies: some tentative calculations, Discussion paper no.69, Statistics Norway.

Atkinson, A.B. and N.H. Stern, 1974, Pigou, taxation and public goods, *Review of Economic Studies* 41, 119-128.

Atkinson, A.B. and J.E. Stiglitz, 1976, The design of tax structure: direct versus indirect taxation, *Journal of Public Economics* 10, 55-75.

Auerbach, A.J., 1985, The theory of excess burden and optimal taxation, in: A.J. Auerbach and M. Feldstein (eds.), *Handbook of Public Economics*, Vol. I, North Holland, Amsterdam.

Ballard, C.L. and S.G. Medema, 1993, The marginal efficiency effects of taxes and subsidies in the presence of externalities: a computational general equilibrium approach, *Journal of Public Economics* 52, 199-216.

Barnett, A.H., 1980, The pigovian tax rule under monopoly, *American Economic Review* 70, 1037-1041.

Barns, D.W., J.A. Edmonds and J.M. Reilly, 1992, Use of the Edmonds-Reilly model to model energy-related greenhouse gas emissions for inclusion in an OECD survey volume, Working paper no. 113, Economics Department, OECD, Paris.

Barrett, S., 1994, Strategic environmental policy and international trade, *Journal of Public Economics* 54, 325-338.

Barro, R.J., 1990, Government spending in a simple model of endogenous growth, *Journal of Political Economy* 98, 103-125.

Barro, R.J. and X. Sala-i-Martin, 1992, Public finance in models of economic growth, *Review of Economic Studies* 59, 645-661.

Baumol, W.J. and W.E. Oates, 1988, *The theory of environmental policy*, Cambridge University Press, Cambridge.

Bayindir-Upmann, T., and M.G. Raith, 1997, Environmental taxation and the double dividend: a drawback for a revenue-neutral tax reform, Working paper No. 274, Institute of Mathematical Economics, University of Bielefeld.

Beaumais O. and L. Ragot, 1998, Tax reform and environmental policy: second-best analysis using a French applied dynamic general equilibrium model, mimeo, University of Metz.

Berndt, E.R. and D.O. Wood, 1979, Engineering and econometric interpretations of energy-capital complementarity, *American Economic Review 69*, 342-354.

Berndt, E.R. and D.M. Hesse, 1986, Measuring and assessing capacity utilization in the manufacturing sectors of nine OECD countries, *European Economic Review* 30, 961-989.

Boadway, R., S. Maital, and M. Prachowny, 1973, Optimal tariffs, optimal taxes and public goods, *Journal of Public Economics* 2, 391- 403.

Bovenberg, A.L., 1995, Environmental taxation and employment, *De Economist* 143, 111 - 140.

Bovenberg, A.L., 1997, Environmental policy, distortionary labour taxation and employment: pollution taxes and the double dividend, in C. Cararro and D. Siniscalco (eds.), *New Directions in the Economic Theory of the Environment*, Cambridge University Press, Cambridge.

Bovenberg, A.L., 1998, Green tax reforms: implications for welfare and distribution, paper presented at the 54th Congress of the IIPF, August 24-28, Cordoba, Argentina.

Bovenberg, A.L. and S. Cnossen, 1991, Fiscal fata morgana (in Dutch), *Economisch Statistische Berichten* 76, 1200-1203.

Bovenberg, A.L. and L.H. Goulder, 1996, Optimal environmental taxation in the presence of other taxes: General equilibrium analyses, *American Economic Review* 86, 985-1000.

Bovenberg, A.L. and G.H.A. van Hagen, 1997, Environmental levies, preexisting tax distortions and distributional concerns, paper presented at the annual conference of the American Economic Association (AEA), New Orleans.

Bovenberg, A.L. and R.A. de Mooij, 1992, Environmental taxation and labor-market distortions, Working paper 9252, CentER for Economic Research, Tilburg University.

Bovenberg, A.L. and R.A. de Mooij, 1994a, Environmental levies and distortionary taxation, *American Economic Review* 94, 1085-89.

Bovenberg, A.L. and R.A. de Mooij, 1994b, Environmental taxes and labor-market distortions, *European Journal of Political Economy* 10, 655-683.

Bovenberg, A.L. and R.A. de Mooij, 1994c, Environmental policy in a small open economy with distortionary labor taxes: a general equilibrium analysis, in: E.C. van Ierland (ed.), *International Environmental Economics*, Elsevier, Amsterdam.

Bovenberg, A.L. and R.A. de Mooij, 1996, Environmental taxation and the double dividend: the role of factor substitution and capital mobility, in: C. Carraro and D. Siniscalco (eds.), *Environmental Fiscal Reform and Unemployment,* Kluwer Academic Publishers, Dordrecht.

Bovenberg, A.L. and R.A. de Mooij, 1997a, Environmental levies and distortionary taxation: Reply, *American Economic Review* 87, 252-253.

Bovenberg, A.L. and F. van der Ploeg, 1994a, Environmental policy, public finance and the labour market in a second-best world, *Journal of Public Economics* 55, 349-390.

Bovenberg, A.L. and F. van der Ploeg, 1994b, Green policies and public finance in a small open economy, *Scandinavian Journal of Economics* 96, 343-363.

Bovenberg, A.L. and F. van der Ploeg, 1994c, Effects of the tax and benefit system on wage formation and unemployment, mimeo, Tilburg University.

Bovenberg, A.L. and F. van der Ploeg, 1996, Optimal taxation, public goods and environmental policy with involuntary unemployment, *Journal of Public Economics* 62, 59-83.

Bovenberg, A.L. and F. van der Ploeg, 1998a, Tax reform, structural unemployment and the environment, *Scandinavian Journal of Economics* 100, 593-610.

Bovenberg, A.L. and F. van der Ploeg, 1998b, Consequences of environmental tax reform for involuntary unemployment and welfare, *Environmental and Resource Economics* (forthcoming).

Bovenberg, A.L. and S. Smulders, 1995, Environmental quality and pollution-saving technological change in a two-sector endogenous growth model, *Journal of Public Economics* 57, 369-391.

Bovenberg, A.L. and S. Smulders, 1996, Transitional impacts of environmental policy in an endogenous growth model, *International Economic Review* 37, 861-893.

Braden, J.B. and C.D. Kolstad, 1991, *Measuring the demand for environmental quality*, Contributions to Economic Analysis no. 198, North Holland, Amsterdam.

Brunello, G., 1996, Labour market institutions and the double dividend hypothesis: An application of the WARM model, in: C. Carraro and D. Siniscalco (eds.), *Environmental Fiscal Reform and Unemployment*, Kluwer Academic Publishers, Dordrecht.

Buchanan, J.M., 1969, External diseconomies, corrective taxes, and market structure, *American Economic Review* 59, 174-177.

Burniaux, J-M., J.P. Martin, G. Nicoletti and J. Oliveira Martins, 1992a, The costs of reducing CO_2 emissions: Evidence from GREEN, Working paper no. 115, Economics Department, OECD, Paris.

Burniaux, J-M., J.P. Martin, G. Nicoletti and J. Oliveira Martins, 1992b, Green -- A multi-sector, multiregion dynamic general equilibrium model for quantifying the costs of curbing CO_2 emissions: A technical manual, Working paper no. 116, Economics Department, OECD, Paris.

Butter, F.A. den, and M.W. Hofkes, 1993, 'Sustainable development with extractive and non-extractive use of the environment in production,' Discussion paper 93-194, Tinbergen Institute, Amsterdam.

Carraro, C., M. Galeotti, and M. Gallo, 1996, Environmental taxation and unemployment: Some evidence on the 'double dividend hypothesis' in Europe, *Journal of Public Economics* 62, 141-181.

Coase, R.H., 1960, The problem of social cost, *Journal of Law and Economics* 3, 1-44.

Conrad, K., and T.F.N. Schmidt, 1997, Double dividend of climate protection and the role of international policy coordination in the EU: an AGE analysis with the GEM-E3 model, Discussion paper no. 97-26, ZEW, University of Mannheim.

Corlett W.J. and D.C. Hague, 1953, Complementarity and the excess burden of taxation, *Review of Economic Studies* 21, 21-30.

CPB, 1992, Long-term economic consequences of energy taxes (in Dutch), Working paper no. 43, The Hague.

CPB, 1993, Effects of an energy tax on small-scale use at low and high energy prices (in Dutch), Working paper no. 64, The Hague.

Cremer, H. F. Gavhari, and N. Ladoux, 1997, On environmental levies and distortionary taxation, mimeo, University of Toulouse.

Cropper M.L. and W.E. Oates, 1992, Environmental economics: a survey, *Journal of Economic Literature* 30, 675 - 740.

Diamond, P.A. and J.A. Mirrlees, 1971, Optimal taxation and public production: I. Production efficiency, *American Economic Review* 61, 8-27.

Draper, D.A.G, 1995, The export market (in Dutch), CPB Research Memorandum 130, The Hague.

European Commission, 1994, Taxation, employment and environment: fiscal reform for reducing unemployment, European Economy no. 56, Part C, Analytical study no. 3, March, Brussels.

European Commission, 1997, Presentation of the new Community system for the taxation of energy products, Commission staff working paper, SEC (97) 1026, Brussels.

Ekins, P. and S. Speck, 1997, Competitiveness and exemptions from environmental taxes in Europe, *Environmental and Resource Economics* (forthcoming).

European Economy, 1992, The climate challenge - economic aspects of limiting CO_2 emissions, Commission of the European Communities, no. 51, September, Brussels.

Ewijk, C. van, and S. van Wijnbergen, 1995, Can abatement overcome the conflict between the environment and economic growth?, *De Economist* 143, 197-216.

Frankhauser, S., 1995, *Valuing climate change: the economics of the greenhouse*, Earthscan, London.

Fuest, C. and B. Huber, 1999, Second-best pollution taxes: an analytical framework and some new results, *Bulletin of Economic Research* 51, 31-38.

Fullerton, D., 1995, Why have separate environmental taxes?, NBER Working Paper 5380, Cambridge, MA.

Fullerton, D., 1997, Environmental levies and distortionary taxation: Comment, *American Economic Review* 87, 245-251.

Fullerton, D. and G.E. Metcalf, 1996, Environmental regulation in a second-best world, mimeo, University of Texas.

Fullerton, D. and G.E. Metcalf, 1997, Environmental taxes and the double-dividend hypothesis: did you really expect something for nothing?, NBER Working Paper 6199, Cambridge, MA.

Goulder, L.H., 1995a, Environmental taxation and the double dividend: A reader's guide, *International Tax and Public Finance* 2, 155-182.

Goulder, L.H., 1995b, Effects of carbon taxes in an economy with prior tax distortions: an intertemporal general equilibrium analysis, *Journal of Environmental Economics and Management* 29, 271-297.

Goulder, L.H., I.W.H. Parry, and D. Burtraw, 1997, Revenue-raising vs. other approaches to environmental protection: the critical significance of preexisting tax distortions, *Rand Journal of Economics* 28, 708-731.

Graafland, J.J. and F. Huizinga, 1997, Estimating a non-linear wage equation for the Netherlands, *CPB Report* 1997/1, 21-25.

Gradus, R., and S. Smulders, 1993, The trade-off between environmental care and long-term growth -- pollution in three prototype growth models, *Journal of Economics* 58, 25-51.

Griffin, J.M., 1981, Engineering and econometric interpretations of energy-capital complementarity: Comment, *American Economic Review* 71, 1100-1104.

Hagen, G.H.A. van, 1995, Tax policy in an open economy with a resource externality and rationing of low-skilled labor, paper presented at the AIO presentation day, Nijmegen, October 27.

Håkonsen, L., 1996, Fiscal and environmental efficiency: a trade-off rather than a double dividend, mimeo, University of Bergen.

Håkonsen, L., 1998, Essays on taxation, efficiency and the environment, Ph.D. Thesis, Norwegian School of Economics and Business Administration.

Hausman, J.A., 1985, Taxes and labor supply, in: A.J. Auerbach and M. Feldstein (eds.), *Handbook of Public Economics*, Vol. I, North Holland, Amsterdam.

Hesse, D.M. and H. Tarkka, 1986, The demand for capital, labor and energy in European manufacturing industry before and after the oil price shocks, *Scandinavian Journal of Economics* 88, 529-46.

Holmlund, B. and A-S. Kolm, 1997, Environmental tax reform in a small open economy with structural unemployment, in: *Economic Studies* 32 (Ph.D. thesis, chapter 3), Uppsala University.

Hoeller, P., A. Dean and J. Nicolaisen, 1991, Macroeconomic implications of reducing greenhouse gas emissions: a survey of empirical studies, *OECD Economic Studies*, No. 16.

Hoeller, P., A. Dean and M. Hayafuji, 1992, New issues, new results: The OECD's second survey of the macroeconomic costs of reducing CO_2 emissions, Working paper no. 123, Economics department, OECD, Paris.

Ilmakunnas, P. and H. Törmä, 1989, Structural change in factor substitution in Finnish manufacturing, *Scandinavian Journal of Economics* 91, 705-721.

Jaffe, A.B., S.R. Peterson, P.R. Portney and R. Stavins, 1995, Environmental regulation and the competitiveness of U.S. manufacturing, *Journal of Economic Literature* 33, 132-164.

Johansson, O., 1995, On externality correcting taxes with distributional considerations in a second-best world, mimeo, Göteborg University.

Keller, W.J., 1980, *Tax incidence: a general equilibrium approach*, North Holland, Amsterdam.

Killinger, S., 1995, Indirect internalization of international environmental externalities, Discussion paper nr. 264, University of Konstanz.

Knoester, A. and N. van der Windt, 1987, Real wages and taxation in ten OECD countries, *Oxford Economic Papers* 49, 151-169.

Koskela, E., and R. Schöb, 1996, Alleviating unemployment: the case for green tax reforms, *European Economic Review* (forthcoming).

Koskela, E., R. Schöb and H-W. Sinn, 1998, Pollution, factor taxation and unemployment, *International Tax and Public Finance* 5, 379-396.

Layard, R., S. Nickell and R. Jackman, 1991, *Unemployment*, Oxford University Press, Oxford.

Lee, D.R. and W.S. Misiolek, 1986, Substituting pollution taxation for general taxation: some implications for efficiency in pollution taxation, *Journal of Environmental Economics and Management* 13, 338-347.

Ligthart, J.E. and F. van der Ploeg, 1995, Environmental quality, public finance and sustainable growth, in Carraro, C. and J. Filer (eds.), *Control and game theoretic models of the environment*, Birkhauser.

Ligthart, J. and F. van der Ploeg, 1997, Environmental policy, tax incidence and the cost of public funds, *Environmental and Resource Economics*, forthcoming.

Lundin, D., 1998, Accounting for geographical residence in a search for welfare-improving carbon dioxide tax reforms, Paper presented at the 54th IIPF congress, Cordoba, Argentina, August 24-27.

Mayeres, I. and S. Proost, 1997, Optimal tax and public investment rules for congestion type of externalities, *Scandinavian Journal of Economics* 99, 261-280.

Mayshar, Y. and S. Yitzhaki, 1995, Dalton-improving indirect tax reform, *American Economic Review* 85, 793-807.

McDonald, I.M. and R.M. Solow, 1981, Wage bargaining and unemployment, *American Economic Review* 71, 896-908.

McGuire, M.C., 1982, Regulation, factor rewards, and international trade, *Journal of Public Economics* 17, 335-354.

Ministry of Economic Affairs, 1989, *Energy in figures*, The Hague.

Mooij, R.A. de, J. van Sinderen and M.W. Gout, 1998, Welfare effects of different public expenditures and taxes in the Netherlands, *Empirica* 25, 263-284.

Mooij, R.A. de, and H.R.J. Vollebergh, 1995, Prospects for European environmental tax reform, in: F.J. Dietz, H.R.J. Vollebergh and J.L. de

Vries (eds.) *Environment, Incentives and the Common Market*, Kluwer Academic Publishers, Dordrecht.

Ng, Y.-K., 1980, Optimal corrective taxes or subsidies when revenue raising imposes an excess burden, *American Economic Review* 70, 744-751.

Nielsen, S.B., L.H. Pedersen and P.B. Sørensen, 1995, Environmental policy, unemployment and endogenous growth, *International Tax and Public Finance* 2, 185-206.

Nordhaus, W.D., 1993, Optimal greenhouse-gas reductions and tax policy in the DICE model, *American Economic Review* 83, 313-317.

Oates, W.E., 1993, Pollution charges as a source of public revenues, in H. Giersch (ed.), *Economic progress and environmental concerns*, Springer Verlag, Berlin.

Oates, W.E., 1995, Green taxes: Can we protect the environment and improve the tax system at the same time?, *Southern Economic Journal* 61, 915-922.

OECD, 1997, *Evaluating Economic Instruments for Environmental Policy*, Paris.

Orosel G.O. and R. Schöb, 1996, Internalizing externalities in second-best tax systems, mimeo, University of Vienna and University of Munich.

Ostro, B., 1994, Estimating the health effects of air pollutants: a method with an application to Jakarta, Rep. No. WPS1301, World Bank, Washington, DC.

Parry, I.H., 1995, Pollution taxes and revenue recycling, *Journal of Environmental Economics and Management* 29, 64-77.

Parry, I.H., 1997, Environmental taxes and quotas in the presence of distortionary taxes in factor markets, *Resource and Energy Economics* 19, 203-220.

Pearce, D.W., 1991, The role of carbon taxes in adjusting to global warming, *Economic Journal* 101, 938-948.

Pezzey, J.C.V, 1992, Sustainability: an interdisciplinary guide, *Environmental Values* 1, 321-362.

Pezzey, J.C.V. and A. Park, 1998, Reflections on the double-dividend debate, *Environmental and Resource Economics*, 3-4, 539-555.

Pigou, A.C., 1932, *The Economics of Welfare*, Fourth Edition, MacMillen & Co, London.

Pindyck, R.S., 1979, Interfuel substitution and the industrial demand for energy: an international comparison, *Review of Economics and Statistics* 61, 169-179.

Pirttilä J. and R. Schöb, 1996, Redistribution and internalization: the many-person Ramsey tax rule revisited, mimeo, University of Helsinki and University of Munich.

Pirttilä J. and M. Tuomala, 1996, Income tax, commodity tax, and environmental policy, mimeo, University of Helsinki and University of Tampere.

Ploeg, F. van der, and A.L. Bovenberg, 1994, Environmental policy, public goods and the marginal cost of public funds, *Economic Journal* 104, 444-454.

Ploeg, F. van der and C. Withagen, 1991, Pollution control and the Ramsey problem, *Environmental and Resource Economics* 1, 215-36.

Poterba, J.M., 1993, Global warming policy: a public finance perspective, *Journal of Economic Perspectives* 7, 47-63.

Proost, S. and D. van Regemorter, 1995, The double dividend and the role of inequality aversion and macroeconomic regimes, *International Tax and Public Finance* 2, 207-219.

Rebelo, S., 1991, Long-run policy analysis and long-run growth, *Journal of Political Economy* 99, 500-521.

Repetto, R. and D. Austin, 1997, *The costs of climate protection: a guide for the perplexed*, World Resource Institute, Washington, D.C.

Repetto, R., R.C. Dower, R. Jenkins and J. Geoghegan, 1992, *Green fees. How a tax shift can work for the environment and the economy*, World Resources Institute, Washington, D.C.

Sandmo, A., 1975, Optimal taxation in the presence of externalities, *Swedish Journal of Economics* 77, 86-98.

Schneider, K., 1997, Involuntary unemployment and environmental policy: the double dividend hypothesis, *Scandinavian Journal of Economics* 99, 45-60.

Schöb, R., 1996, Evaluating tax reforms in the presence of externalities, *Oxford Economic Papers* 48, 537-555.

Schöb, R., 1997, Environmental taxes and pre-existing distortions: the normalization trap, *International Tax and Public Finance* 4, 167-176.

Schwartz, J. and R. Repetto, 1997, Nonseparable utility and the double dividend debate: reconsidering the tax-interaction effect, mimeo University of Maryland and World Resource Institute, Washington, D.C..

Siebert, H., J. Eichenberger, R. Gronych and R. Pethig, 1980, *Trade and environment: A theoretical inquiry*, Elsevier, Dordrecht.

Smith, S, 1992, Taxation and the environment: a survey, *Fiscal Studies* 13, 21-57.

Smulders, S., and R. Gradus, 1996, Pollution abatement and long-term growth, *European Journal of Political Economy* 12, 505-532.

Smulders, S., and P. Sen, 1998, The double dividend hypothesis and trade liberalization, paper presented at the NAKE Research Day, Amsterdam, October 23.

Standaert, S., 1992, The macro-sectoral effects of an EC-wide energy tax: simulation experiments for 1993-2005, *European Economy*, special edition no. 1: The economics of limiting CO_2 emissions, Brussels.

Tahvonen, O. and J. Kuuluvainen, 1993, Economic growth, pollution and renewable resources, *Journal of Environmental Economics and Management* 24, 101-118.

Tang, P.J.G., R.A. de Mooij and R. Nahuis, 1998, An economic evaluation of alternative approaches for limiting the costs of unilateral action to slow down global climate change: simulations with WorldScan, *European Economy* 1998 no. 1, pp. 11-35, DG II European Commission, Brussels.

Terkla, D., 1984, The efficiency value of effluent tax revenues, *Journal of Environmental Economics and Management* 11, 107-123.

Tol, R.S.J., 1995, The damage costs of climate change: towards more comprehensive calculations, *Environmental and Resource Economics* 5, 353-374.

Topham, N., 1984, A reappraisal and recalculation of the marginal cost of public funds, *Public Finance* 3, 394-405.

Tullock, G., 1967, Excess benefit, *Water Resources Research* 3, 643-644.

Whalley, J. and R. Wigle, 1991, The international incidence of carbon taxes, in R. Dornbusch and J.M. Poterba (eds.), *Global Warming: Economic Policy Responses*, MIT Press, Cambridge, MA.

Wildasin, D.E., 1984, On public good provision with distortionary taxation, *Economic Enquiry* 22, 227-243.

Williams, R.C., 1998a, Revisiting the cost of protectionism: the role of tax distortions in the labor market, *Journal of International Economics*, forthcoming.

Williams, R.C., 1998b, The impact of pre-existing taxes on the benefits of regulations affecting health or productivity, mimeo Stanford University.

WRR, 1992, Environmental policy: Strategy, instruments and enforcement (in Dutch), Government Report no. 41, SDU, The Hague.

Zuidema, T. and A. Nentjes, 1997, Health damage of air pollution: An estimate of dose-response relationship for the Netherlands, *Environmental and Resource Economics* 9, 291-308.